Principles of Steel Making

Principles of Steel Making

Edited by
Greg Mayhew

Larsen & Keller
www.larsen-keller.com

Principles of Steel Making
Edited by Greg Mayhew
ISBN: 978-1-63549-703-8 (Hardback)

Larsen & Keller
Published by Larsen and Keller Education,
5 Penn Plaza,
19th Floor,
New York, NY 10001, USA

Cataloging-in-Publication Data

Principles of steel making / edited by Greg Mayhew.
p. cm.
Includes bibliographical references and index.
ISBN 978-1-63549-703-8
1. Steel. 2. Steel--Metallurgy. 3. Steel founding. I. Mayhew, Greg.
TS305 .P75 2018
669.142--dc23

Printed and bound in China.

For more information regarding Larsen and Keller Education and its products, please visit the publisher's website www.larsen-keller.com

Table of Contents

Preface

Steel making refers to the process of removing impurities in the form of phosphorus, silicon, carbon, nitrogen, etc. from raw steel and adding alloy elements like chromium, manganese, nickel, etc. to produce different grades of steel. The three main processes included in this field are basic oxygen steelmaking, ladle (metallurgy), and HIsarna steelmaking process. Most of the topics introduced in this book cover new methods and the applications of steel making. This textbook is an essential guide for both academicians and those who wish to pursue this discipline further.

Given below is the chapter wise description of the book:

Chapter 1- Steel is an alloy of iron, carbon and other elements. It is an important material in engineering as it is cheap and has a high tensile strength. The amount of carbon in the alloy and the form in which it is present regulates steel's characteristics. The chapter on steel and steel making offers an insightful focus, keeping in mind the complex subject matter.

Chapter 2- Stainless steel is a form of steel that does not easily rust or corrode in presence of water as normal steel does. Stainless steel can be classified into three types based on their crystalline structure. They are austenitic, martensitic and ferritic. This chapter is an overview of the subject matter incorporating all the major aspects of stainless steel.

Chapter 3- There has been a growing need to have ecologically-viable technology to produce steel in terms of available resources, energy consumption and impact on environment. Two methods that follow these principles are energy optimizing furnace and CONARC. This chapter discusses in detail the theories and methodologies related to modern steelmaking practices.

Chapter 4- The industrial plants that manufacture steel are called steel mills. They also design casting-products such as blooms, billets and slabs. A ladle in a foundry is a vessel that pours molten steel. Casting ladle, treatment ladle, and transfer ladle are three types of ladles based on their respective uses. The chapter closely examines the key concepts of steel mills and ladle metallurgy to provide an extensive understanding of the subject.

Chapter 5- After the removal of superheat and latent heat, liquid changes into solid. This transformation process is called solidification. While pure metals solidify at a particular temperature, alloys can solidify at different temperature. The process depends on the temperature gradient, under cooling, growth rate, and alloy constitution. This section also explores the concept of casting by which a liquid material is poured into a mold and left to solidify. All the diverse principles of solidification and casting have been carefully analyzed in this chapter.

Chapter 6- During the process of making steel, there are many impurities present that need to be removed. Impurities such as carbon, silicon, phosphorus, manganese and sulphur are removed through oxidation. After the process, oxygen may still be left. This is removed through deoxidization. This chapter has been carefully written to provide an easy understanding of the varied facets of steelmaking reactions.

At the end, I would like to thank all those who dedicated their time and efforts for the successful completion of this book. I also wish to convey my gratitude towards my friends and family who supported me at every step.

Editor

1

Steel and Steelmaking

Steel is an alloy of iron, carbon and other elements. It is an important material in engineering as it is cheap and has a high tensile strength. The amount of carbon in the alloy and the form in which it is present regulates steel's characteristics. The chapter on steel and steel making offers an insightful focus, keeping in mind the complex subject matter.

Steel

The steel cable of a colliery winding tower

Steel is an alloy of iron and other elements, primarily carbon, that is widely used in construction and other applications because of its high tensile strength and low cost. Steel's base metal is iron, which is able to take on two crystalline forms (allotropic forms), body centered cubic and face centered cubic (FCC), depending on its temperature. It is the interaction of those allotropes with the alloying elements, primarily carbon, that gives steel and cast iron their range of unique properties. In the body-centred cubic arrangement, there is an iron atom in the centre of each cube, and in the face-centred cubic, there is one at the center of each of the six faces of the cube. Carbon, other elements, and inclusions within iron act as hardening agents that prevent the movement of dislocations that otherwise occur in the crystal lattices of iron atoms.

The carbon in typical steel alloys may contribute up to 2.1% of its weight. Varying the amount of alloying elements, their presence in the steel either as solute elements, or as precipitated phases, retards the movement of those dislocations that make iron comparatively ductile and weak, and thus controls its qualities such as the hardness, ductility, and tensile strength of the resulting steel.

Steel's strength compared to pure iron is only possible at the expense of iron's ductility, of which iron has an excess.

Steel was produced in bloomery furnaces for thousands of years, but its extensive use began after more efficient production methods were devised in the 17th century, with the production of blister steel and then crucible steel. With the invention of the Bessemer process in the mid-19th century, a new era of mass-produced steel began. This was followed by Siemens-Martin process and then Gilchrist-Thomas process that refined the quality of steel. With their introductions, mild steel replaced wrought iron.

Further refinements in the process, such as basic oxygen steelmaking (BOS), largely replaced earlier methods by further lowering the cost of production and increasing the quality of the product. Today, steel is one of the most common materials in the world, with more than 1.3 billion tons produced annually. It is a major component in buildings, infrastructure, tools, ships, automobiles, machines, appliances, and weapons. Modern steel is generally identified by various grades defined by assorted standards organizations.

Definitions and Related Materials

The noun *steel* originates from the Proto-Germanic adjective *stahliją* or *stakhlijan* (*made of steel*), which is related to *stahlaz* or *stahliją(standing firm)*.

The carbon content of steel is between 0.002% and 2.1% by weight for plain iron–carbon alloys. These values vary depending on alloying elements such as manganese, chromium, nickel, iron, tungsten, carbon and so on. Basically, steel is an iron-carbon alloy that does not undergo eutectic reaction. In contrast, cast iron does undergo eutectic reaction. Too little carbon content leaves (pure) iron quite soft, ductile, and weak. Carbon contents higher than those of steel make an alloy, commonly called pig iron, that is brittle (not malleable). While iron alloyed with carbon is called carbon steel, alloy steel is steel to which other alloying elements have been intentionally added to modify the characteristics of steel. Common alloying elements include: manganese, nickel, chromium, molybdenum, boron, titanium, vanadium, tungsten, cobalt, and niobium. Additional elements are also important in steel: phosphorus, sulfur, silicon, and traces of oxygen, nitrogen, and copper, that are most frequently considered undesirable.

Alloys with a higher than 2.1% carbon content, depending on other element content and possibly on processing, are known as cast iron. Cast iron is not malleable even when hot, but it can be formed by casting as it has a lower melting point than steel and good castability properties. Certain compositions of cast iron, while retaining the economies of melting and casting, can be heat treated after casting to make malleable iron or ductile iron objects. Steel is also distinguishable from wrought iron (now largely obsolete), which may contain a small amount of carbon but large amounts of slag.

Material Properties

Iron is commonly found in the Earth's crust in the form of an ore, usually an iron oxide, such as magnetite, hematite etc. Iron is extracted from iron ore by removing the oxygen through its combination with a preferred chemical partner such as carbon that is then lost to the atmosphere as carbon dioxide. This process, known as smelting, was first applied to metals with lower melting points, such as tin, which melts at about 250 °C (482 °F) and copper, which melts at about 1,100 °C (2,010 °F) and the combination, bronze, which is liquid at less than 1,083 °C (1,981 °F). In comparison, cast iron melts at about 1,375 °C (2,507 °F). Small quantities of iron were smelted in ancient times, in the solid state, by heating the ore in a charcoal fire and then welding the clumps together with a hammer and in the process squeezing out the impurities. With care, the carbon content could be controlled by moving it around in the fire. Unlike copper and tin, liquid or solid iron dissolves carbon quite readily.

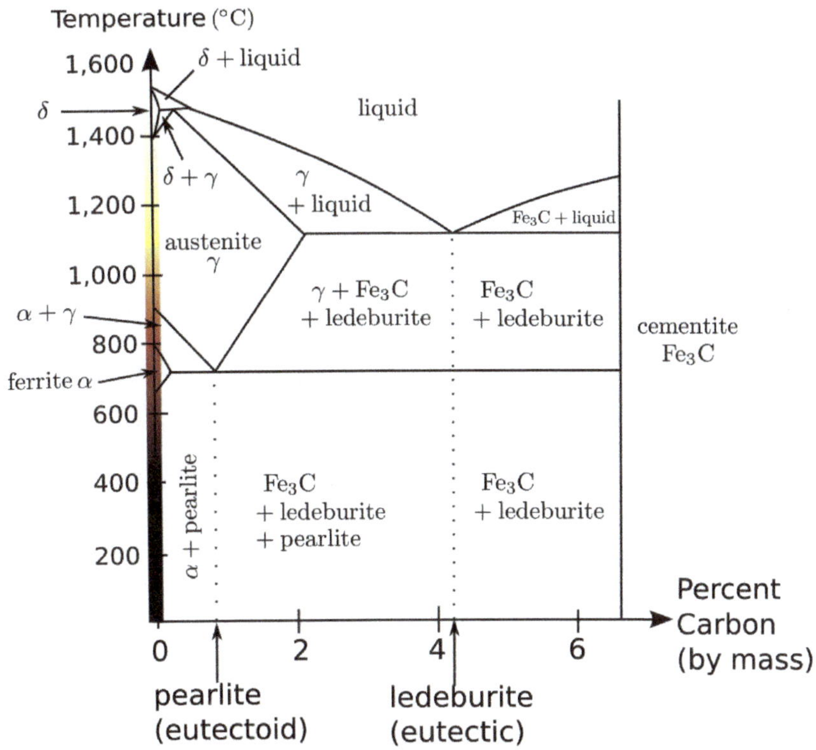

Iron-carbon phase diagram, showing the conditions necessary to form different phases

All of these temperatures could be reached with ancient methods used since the Bronze Age. Since the oxidation rate of iron increases rapidly beyond 800 °C (1,470 °F), it is important that smelting take place in a low-oxygen environment. Smelting, using carbon to reduce iron oxides, results in an alloy (pig iron) that retains too much carbon to be called steel. The excess carbon and other impurities are removed in a subsequent step.

Other materials are often added to the iron/carbon mixture to produce steel with desired properties. Nickel and manganese in steel add to its tensile strength and make the austenite form of the iron-carbon solution more stable, chromium increases hardness and melting temperature, and vanadium also increases hardness while making it less prone to metal fatigue.

To inhibit corrosion, at least 11% chromium is added to steel so that a hard oxide forms on the metal surface; this is known as stainless steel. Tungsten slows the formation of cementite, keeping carbon in the iron matrix and allowing martensite to preferentially form at slower quench rates, resulting in high speed steel. On the other hand, sulfur, nitrogen, and phosphorus are considered contaminants that make steel more brittle and are removed from the steel melt during processing.

The density of steel varies based on the alloying constituents but usually ranges between 7,750 and 8,050 kg/m³ (484 and 503 lb/cu ft), or 7.75 and 8.05 g/cm³ (4.48 and 4.65 oz/cu in).

Even in a narrow range of concentrations of mixtures of carbon and iron that make a steel, a number of different metallurgical structures, with very different properties can form. Understanding such properties is essential to making quality steel. At room temperature, the most stable form of pure iron is the body-centered cubic (BCC) structure called alpha iron or α-iron. It is a fairly soft metal that can dissolve only a small concentration of carbon, no more than 0.005% at 0 °C (32 °F) and 0.021 wt% at 723 °C (1,333 °F). The inclusion of carbon in alpha iron is called ferrite. At 910 °C pure iron transforms into a face-centered cubic (FCC) structure, called gamma iron or γ-iron. The inclusion of carbon in gamma iron is called austenite. The more open FCC structure of austenite can dissolve considerably more carbon, as much as 2.1% (38 times that of ferrite) carbon at 1,148 °C (2,098 °F), which reflects the upper carbon content of steel, beyond which is cast iron. When carbon moves out of solution with iron it forms a very hard, but brittle material called cementite (Fe_3C).

When steels with exactly 0.8% carbon (known as a eutectoid steel), are cooled, the austenitic phase (FCC) of the mixture attempts to revert to the ferrite phase (BCC). The carbon no longer fits within the FCC austenite structure, resulting in an excess of carbon. One way for carbon to leave the austenite is for it to precipitate out of solution as cementite, leaving behind a surrounding phase of BCC iron called ferrite with a small percentage of carbon in solution. The two, ferrite and cementite, precipitate simultaneously producing a layered structure called pearlite, named for its resemblance to mother of pearl. In a hypereutectoid composition (greater than 0.8% carbon), the carbon will first precipitate out as large inclusions of cementite at the austenite grain boundaries until the percenage of carbon in the grains has decreased to the eutectoid composition (0.8% carbon), at which point the pearlite structure forms. For steels that have less than 0.8% carbon (hypoeutectoid), ferrite will first form within the grains until the remaining composition rises to 0.8% of carbon, at which point the pearlite structure will form. No large inclusions of cementite will form at the boundaries in hypoeuctoid steel. The above assumes that the cooling process is very slow, allowing enough time for the carbon to migrate.

As the rate of cooling is increased the carbon will have less time to migrate to form carbide at the grain boundaries but will have increasingly large amounts of pearlite of a finer and finer structure within the grains; hence the carbide is more widely dispersed and acts to prevent slip of defects within those grains, resulting in hardening of the steel. At the very high cooling rates produced by quenching, the carbon has no time to migrate but is locked within the face center austenite and forms martensite. Martensite is a highly strained and stressed, supersaturated form of carbon and iron and is exceedingly hard but brittle. Depending on the carbon content, the martensitic phase takes different forms. Below 0.2% carbon, it takes on a ferrite BCC crystal form, but at higher carbon content it takes a body-centered tetragonal (BCT) structure. There is no thermal activation energy for the transformation from austenite to martensite. Moreover, there is no compositional change so the atoms generally retain their same neighbors.

Martensite has a lower density (it expands during the cooling) than does austenite, so that the transformation between them results in a change of volume. In this case, expansion occurs. Internal stresses from this expansion generally take the form of compression on the crystals of martensite and tension on the remaining ferrite, with a fair amount of shear on both constituents. If quenching is done improperly, the internal stresses can cause a part to shatter as it cools. At the very least, they cause internal work hardening and other microscopic imperfections. It is common for quench cracks to form when steel is water quenched, although they may not always be visible.

Heat Treatment

There are many types of heat treating processes available to steel. The most common are annealing, quenching, and tempering. Heat treatment is effective on compositions above the eutectoid composition (hypereutectoid) of 0.8% carbon. Hypoeutectoid steel does not benefit from heat treatment.

Annealing is the process of heating the steel to a sufficiently high temperature to relieve local internal stresses. It does not create a general softening of the product but only locally relieves strains and stresses locked up within the material. Annealing goes through three phases: recovery, recrystallization, and grain growth. The temperature required to anneal a particular steel depends on the type of annealing to be achieved and the alloying constituents.

Quenching involves heating the steel to create the austenite phase then quenching it in water or oil. This rapid cooling results in a hard but brittle martensitic structure. The steel is then tempered, which is just a specialized type of annealing, to reduce brittleness. In this application the annealing (tempering) process transforms some of the martensite into cementite, or spheroidite and hence it reduces the internal stresses and defects. The result is a more ductile and fracture-resistant steel.

Steel Production

Iron ore pellets for the production of steel

When iron is smelted from its ore, it contains more carbon than is desirable. To become steel, it must be reprocessed to reduce the carbon to the correct amount, at which point other elements can be added. In the past, steel facilities would cast the raw steel product into ingots which would be stored until use in further refinement processes that resulted in the finished product. In modern facilities, the initial product is close to the final composition and is continuously cast into long slabs, cut and shaped into bars and extrusions and heat treated to produce a final product. Today only a small fraction is cast into ingots. Approximately 96% of steel is continuously cast, while only 4% is produced as ingots.

The ingots are then heated in a soaking pit and hot rolled into slabs, billets, or blooms. Slabs are hot or cold rolled into sheet metal or plates. Billets are hot or cold rolled into bars, rods, and wire. Blooms are hot or cold rolled into structural steel, such as I-beams and rails. In modern steel mills these processes often occur in one assembly line, with ore coming in and finished steel products coming out. Sometimes after a steel's final rolling it is heat treated for strength, however this is relatively rare.

History of Steelmaking

Bloomery smelting during the Middle Ages

Ancient Steel

Steel was known in antiquity, and possibly was produced in bloomeries and crucibles.

The earliest known production of steel are pieces of ironware excavated from an archaeological site in Anatolia (Kaman-Kalehoyuk) and are nearly 4,000 years old, dating from 1800 BC. Horace identifies steel weapons like the *falcata* in the Iberian Peninsula, while Noric steel was used by the Roman military.

The reputation of *Seric iron* of South India (wootz steel) amongst the rest of the world grew considerably. South Indian and Mediterranean sources including Alexander the Great (3rd c. BC) recount the presentation and export to the Greeks of 100 talents worth of such steel. Metal production sites in Sri Lanka employed wind furnaces driven by the monsoon winds, capable of producing high-carbon steel. Large-scale Wootz steel production in Tamilakam using crucibles and carbon sources such as the plant Avāram occurred by the sixth century BC, the pioneering precursor to modern steel production and metallurgy.

The Chinese of the Warring States period (403–221 BC) had quench-hardened steel, while Chinese of the Han dynasty (202 BC – 220 AD) created steel by melting together wrought iron with cast iron, gaining an ultimate product of a carbon-intermediate steel by the 1st century AD. The Haya people of East Africa invented a type of furnace they used to make carbon steel at 1,802 °C (3,276 °F) nearly 2,000 years ago. East African steel has been suggested by Richard Hooker to date back to 1400 BC.

Wootz Steel and Damascus Steel

Evidence of the earliest production of high carbon steel in the Indian Subcontinent are found in Kodumanal in Tamil Nadu area, Golconda in Andhra Pradesh area and Karnataka, and in Samanalawewa areas of Sri Lanka. This came to be known as Wootz steel, produced in South India by about sixth century BC and exported globally. The steel technology existed prior to 326 BC in the region as they are mentioned in literature of Sangam Tamil, Arabic and Latin as the finest steel in the world exported to the Romans, Egyptian, Chinese and Arab worlds at that time – what they called *Seric Iron*. A 200 BC Tamil trade guild in Tissamaharama, in the South East of Sri Lanka, brought with them some of the oldest iron and steel artifacts and production processes to the island from the classical period. The Chinese and locals in Anuradhapura, Sri Lanka had also adopted the production methods of creating Wootz steel from the Chera Dynasty Tamils of South India by the 5th century AD. In Sri Lanka, this early steel-making method employed a unique wind furnace, driven by the monsoon winds, capable of producing high-carbon steel. Since the technology was acquired from the Tamilians from South India, the origin of steel technology in India can be conservatively estimated at 400–500 BC.

The manufacture of what came to be called Wootz, or Damascus steel, famous for its durability and ability to hold an edge, may have been taken by the Arabs from Persia, who took it from India. It was originally created from a number of different materials including various trace elements, apparently ultimately from the writings of Zosimos of Panopolis. In 327 BCE, Alexander the Great was rewarded by the defeated King Porus, not with gold or silver but with 30 pounds of steel. Recent studies have suggested that carbon nanotubes were included in its structure, which might explain some of its legendary qualities, though given the technology of that time, such qualities were produced by chance rather than by design. Natural wind was used where the soil containing iron was heated by the use of wood. The ancient Sinhalese managed to extract a ton of steel for every 2 tons of soil, a remarkable feat at the time. One such furnace was found in Samanalawewa and archaeologists were able to produce steel as the ancients did.

Crucible steel, formed by slowly heating and cooling pure iron and carbon (typically in the form of charcoal) in a crucible, was produced in Merv by the 9th to 10th century AD. In the 11th century, there is evidence of the production of steel in Song China using two techniques: a "berganesque" method that produced inferior, inhomogeneous, steel, and a precursor to the modern Bessemer process that used partial decarbonization via repeated forging under a cold blast.

Modern Steelmaking

A Bessemer converter in Sheffield, England

Since the 17th century the first step in European steel production has been the smelting of iron ore into pig iron in a blast furnace. Originally employing charcoal, modern methods use coke, which has proven more economical.

Processes Starting from Bar Iron

In these processes pig iron was refined (fined) in a finery forge to produce bar iron, which was then used in steel-making.

The production of steel by the cementation process was described in a treatise published in Prague in 1574 and was in use in Nuremberg from 1601. A similar process for case hardening armour and files was described in a book published in Naples in 1589. The process was introduced to England in about 1614 and used to produce such steel by Sir Basil Brooke at Coalbrookdale during the 1610s.

The raw material for this process were bars of iron. During the 17th century it was realized that the best steel came from oregrounds iron of a region north of Stockholm, Sweden. This was still the usual raw material source in the 19th century, almost as long as the process was used.

Crucible steel is steel that has been melted in a crucible rather than having been forged, with the result that it is more homogeneous. Most previous furnaces could not reach high enough temperatures to melt the steel. The early modern crucible steel industry resulted from the invention of Benjamin Huntsman in the 1740s. Blister steel (made as above) was melted in a crucible or in a furnace, and cast (usually) into ingots.

Processes Starting from Pig Iron

The modern era in steelmaking began with the introduction of Henry Bessemer's Bessemer process in 1855, the raw material for which was pig iron. His method let him produce steel in large quantities cheaply, thus mild steel came to be used for most purposes for which wrought iron was formerly used. The Gilchrist-Thomas process (or *basic Bessemer process*) was an improvement to the Bessemer process, made by lining the converter with a basic material to remove phosphorus.

A Siemens-Martin steel oven from the Brandenburg Museum of Industry.

White-hot steel pouring out of an electric arc furnace.

Another 19th-century steelmaking process was the Siemens-Martin process, which complemented the Bessemer process. It consisted of co-melting bar iron (or steel scrap) with pig iron.

These methods of steel production were rendered obsolete by the Linz-Donawitz process of basic oxygen steelmaking (BOS), developed in the 1950s, and other oxygen steel making methods. Basic oxygen steelmaking is superior to previous steelmaking methods because the oxygen pumped into the furnace limited impurities, primarily nitrogen, that previously had entered from the air used. Today, electric arc furnaces (EAF) are a common method of reprocessing scrap metal to create new steel. They can also be used for converting pig iron to steel, but they use a lot of electrical energy (about 440 kWh per metric ton), and are thus generally only economical when there is a plentiful supply of cheap electricity.

Steel Industry

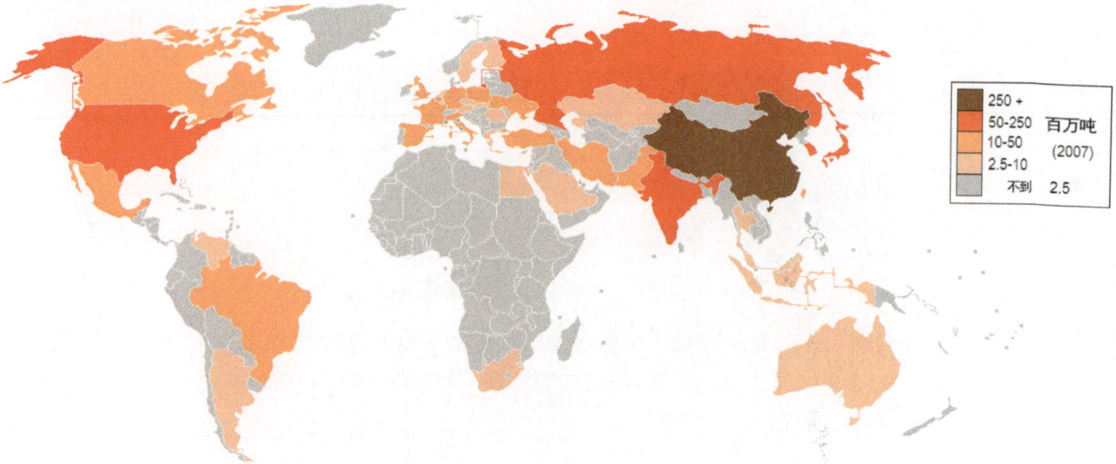

Steel production (in million tons) by country in 2007

A steel plant in the United Kingdom.

It is common today to talk about "the iron and steel industry" as if it were a single entity, but historically they were separate products. The steel industry is often considered an indicator of economic progress, because of the critical role played by steel in infrastructural and overall economic development.

In 1980, there were more than 500,000 U.S. steelworkers. By 2000, the number of steelworkers fell to 224,000.

The economic boom in China and India has caused a massive increase in the demand for steel in recent years. Between 2000 and 2005, world steel demand increased by 6%. Since 2000, several Indian and Chinese steel firms have risen to prominence, such as Tata Steel (which bought Corus Group in 2007), Baosteel Group and Shagang Group. ArcelorMittal is however the world's largest steel producer.

In 2005, the British Geological Survey stated China was the top steel producer with about one-third of the world share; Japan, Russia, and the US followed respectively.

In 2008, steel began trading as a commodity on the London Metal Exchange. At the end of 2008, the steel industry faced a sharp downturn that led to many cut-backs.

The world steel industry peaked in 2007. That year, ThyssenKrupp spent $12 billion to build the two most modern mills in the world, in Calvert, Alabama and Sepetiba, Rio de Janeiro, Brazil. The worldwide Great Recession starting in 2008, however, sharply lowered demand and new construction, and so prices fell. ThyssenKrupp lost $11 billion on its two new plants, which sold steel below the cost of production.

Recycling

Steel is one of the world's most-recycled materials, with a recycling rate of over 60% globally; in the United States alone, over 82,000,000 metric tons (81,000,000 long tons) was recycled in the year 2008, for an overall recycling rate of 83%.

Contemporary Steel

Bethlehem Steel in Bethlehem, Pennsylvania, was one of the world's largest manufacturers of steel before its closure in 2003 and later conversion into a casino.

Carbon Steels

Modern steels are made with varying combinations of alloy metals to fulfill many purposes. Carbon steel, composed simply of iron and carbon, accounts for 90% of steel production. Low alloy steel is alloyed with other elements, usually molybdenum, manganese, chromium, or nickel, in amounts of up to 10% by weight to improve the hardenability of thick sections. High strength low alloy steel has small additions (usually < 2% by weight) of other elements, typically 1.5% manganese, to provide additional strength for a modest price increase.

Recent Corporate Average Fuel Economy (CAFE) regulations have given rise to a new variety of steel known as Advanced High Strength Steel (AHSS). This material is both strong and ductile so that vehicle structures can maintain their current safety levels while using less material. There are several commercially available grades of AHSS, such as dual-phase steel, which is heat treated to contain both a ferritic and martensitic microstructure to produce a formable, high strength steel. Transformation Induced Plasticity (TRIP) steel involves special alloying and heat treatments to stabilize amounts of austenite at room temperature in normally austenite-free low-alloy ferritic steels. By applying strain, the austenite undergoes a phase transition to martensite without the addition of heat. Twinning Induced Plasticity (TWIP) steel uses a specific type of strain to increase the effectiveness of work hardening on the alloy.

Carbon Steels are often galvanized, through hot-dip or electroplating in zinc for protection against rust.

Alloy Steels

Stainless steels contain a minimum of 11% chromium, often combined with nickel, to resist corrosion. Some stainless steels, such as the ferritic stainless steels are magnetic, while others, such as the austenitic, are nonmagnetic. Corrosion-resistant steels are abbreviated as CRES.

Some more modern steels include tool steels, which are alloyed with large amounts of tungsten and cobalt or other elements to maximize solution hardening. This also allows the use of precipitation hardening and improves the alloy's temperature resistance. Tool steel is generally used in axes, drills, and other devices that need a sharp, long-lasting cutting edge. Other special-purpose alloys include weathering steels such as Cor-ten, which weather by acquiring a stable, rusted surface, and so can be used un-painted. Maraging steel is alloyed with nickel and other elements, but unlike most steel contains little carbon (0.01%). This creates a very strong but still malleable steel.

Eglin steel uses a combination of over a dozen different elements in varying amounts to create a relatively low-cost steel for use in bunker buster weapons. Hadfield steel (after Sir Robert Hadfield) or manganese steel contains 12–14% manganese which when abraded strain-hardens to form an incredibly hard skin which resists wearing. Examples include tank tracks, bulldozer blade edges and cutting blades on the jaws of life.

In 2016 a breakthrough in creating a strong light aluminium steel alloy which might be suitable in applications such as aircraft was announced by researchers at Pohang University of Science and Technology. Adding small amounts of nickel was found to result in precipitation as nano particles of brittle B2 intermetallic compounds which had previously resulted in weakness. The result was a cheap strong light steel alloy—nearly as strong as titanium at ten percent the cost—which is slated for trial production at industrial scale by POSCO, a Korean steelmaker.

Standards

Most of the more commonly used steel alloys are categorized into various grades by standards organizations. For example, the Society of Automotive Engineers has a series of grades defining many types of steel. The American Society for Testing and Materials has a separate set of standards, which define alloys such as A36 steel, the most commonly used structural steel in the United States. The JIS also define series of steel grades that are being used extensively in Japan as well in third world countries.

Uses

Iron and steel are used widely in the construction of roads, railways, other infrastructure, appliances, and buildings. Most large modern structures, such as stadiums and skyscrapers, bridges, and airports, are supported by a steel skeleton. Even those with a concrete structure employ steel for reinforcing. In addition, it sees widespread use in major appliances and cars. Despite growth in usage of aluminium, it is still the main material for car bodies. Steel is used in a variety of other construction materials, such as bolts, nails, and screws and other household products and cooking utensils.

A roll of steel wool

Other common applications include shipbuilding, pipelines, mining, offshore construction, aerospace, white goods (e.g. washing machines), heavy equipment such as bulldozers, office furniture, steel wool, tools, and armour in the form of personal vests or vehicle armour (better known as rolled homogeneous armour in this role).

Historical

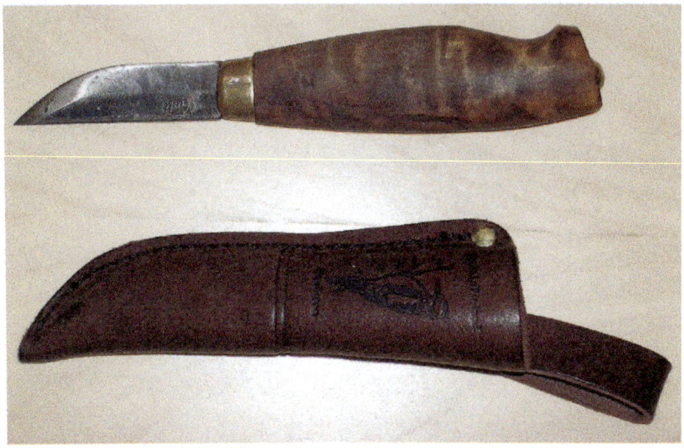

A carbon steel knife

Before the introduction of the Bessemer process and other modern production techniques, steel was expensive and was only used where no cheaper alternative existed, particularly for the cutting edge of knives, razors, swords, and other items where a hard, sharp edge was needed. It was also used for springs, including those used in clocks and watches.

With the advent of speedier and thriftier production methods, steel has become easier to obtain and much cheaper. It has replaced wrought iron for a multitude of purposes. However, the availability of plastics in the latter part of the 20th century allowed these materials to replace steel in some applications due to their lower fabrication cost and weight. Carbon fiber is replacing steel in some cost insensitive applications such as aircraft, sports equipment and high end automobiles.

Long Steel

A steel bridge

A steel pylon suspending overhead power lines

- As reinforcing bars and mesh in reinforced concrete
- Railroad tracks
- Structural steel in modern buildings and bridges
- Wires
- Input to reforging applications

Flat carbon Steel

- Major appliances
- Magnetic cores
- The inside and outside body of automobiles, trains, and ships.

Weathering Steel (COR-TEN)

- Intermodal containers
- Outdoor sculptures
- Architecture
- Highliner train cars

Stainless Steel

A stainless steel gravy boat

- Cutlery
- Rulers
- Surgical instruments
- Watches
- Guns
- Rail passenger vehicles
- Tablets
- Trash Cans
- Body piercing jewellery

Low-background Steel

Steel manufactured after World War II became contaminated with radionuclides by nuclear weapons testing. Low-background steel, steel manufactured prior to 1945, is used for certain radiation-sensitive applications such as Geiger counters and radiation shielding.

Modern Steelmaking

The concept of modern steelmaking is to make use of the steelmaking vessels like converter, ladle and tundish of a continuous caster. In all these vessels molten steel is handled for one or the other purpose. For examples ladles are used to transfer the molten steel either to ingot casting or continuous casting. Tundish of a continuous caster is used to transfer molten steel to the continuous casting mould. In all these vessels the residence time of molten steel is sufficiently long so as to carry out some refining operations like composition adjustment, removal of gases, control of S, removal of inclusions etc. in ladle and tundish. This has led into the development of ladles, tundishes for some refining operations like deoxidation, inclusion modification, desulphurization etc. and other operations like composition adjustment, inclusion removal etc. The basic idea of employing ladles and tundishes for either refining or composition adjustment or for producing clean steels is to use the steelmaking units like converter and electric furnace for producing steels without much bothering for final chemistry. Modern steelmaking comprises of hot metal / scrap to finished products through the following

a) Primary steelmaking

b) Secondary steelmaking

c) Continuous casting

d) Finishing operations

Primary Steelmaking

Primary steelmaking consists of refining of hot metal or scrap +hot metal to steel in a) converter and b) Electric furnace. The objective is to refine hot metal to the nearly desired chemistry.

a) Types of converter steelmaking

- In converter steelmaking pure oxygen is blown from top through a water cooled lance fitted with multi-hole nozzles. This technology of refining of hot metal is called top blown steelmaking.
- In another version of converter steelmaking oxygen is blown from top and bath is gas stirred through the bottom. These are called combined top blowing and bottom stirred processes.
- In some converters, O_2 is blown from top and bottom and these processes are called top and bottom blowing, Duplex blowing or hybrid blowing.

- In some converters oxygen is blown through the bottom and the process is bottom blown converter. This variant is not popular amongst steelmakers.

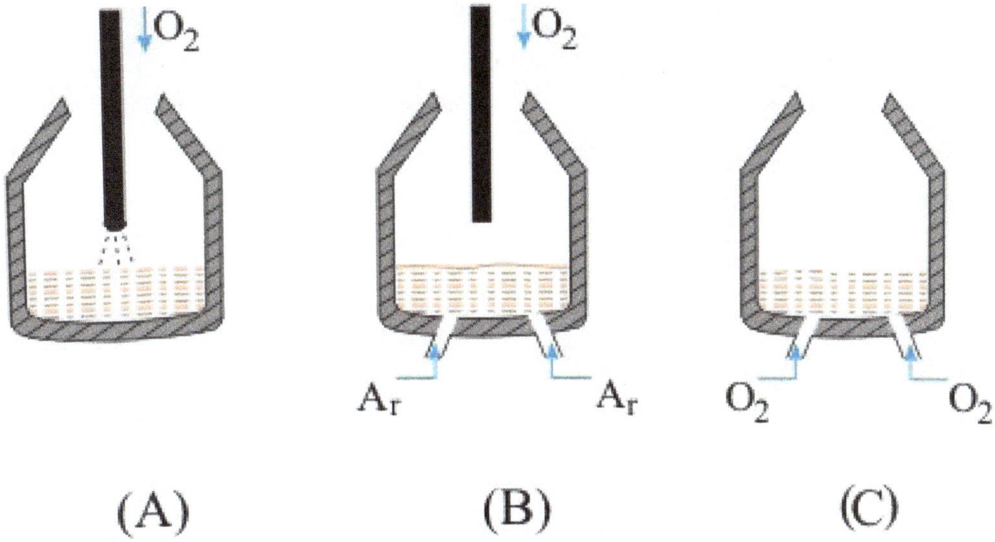

Types of converter steelmaking (a)Top blown steelmaking
(b) Combined top and bottom blowing, and (c) Bottom blowing

It is important to note that in all different types of converter steelmaking practices, a pear shaped vessel is used and blast furnace hot metal is refined to plain carbon steel. Some amount of scrap is also used.

Principle Chemical Reactions

Hot metal contains $C \sim 3.5$ to 4%, $Si \sim 0.6$ to 1% $Mn \sim 0.6$ to 0.8% and $P \sim 0.1$ to 0.2%. Oxygen is blown from top and the following reactions occur:

$[Fe]+[O]=(FeO)$	1
$[C]+[O]=\{CO\}$	2
$[Si]+2[O]=(SiO_2)$	3
$[Mn]+[O]=(MnO)$	4
$2[P]+5[O]=(P_2O_5)$	5
$[C]+(FeO)=\{CO\}+[Fe]$	6
$(Fe)+(MnO)=(FeO)+[Mn]$	7

Note the following:

- No heat is supplied from outside. The heat produced due to chemical reactions is sufficient enough to raise the temperature of hot metal from around $1250^{\circ}C$ to 1300° to molten steel tapping temperature of $1600^{\circ}C$ to $1650^{\circ}C$.

- Except carbon which is removed as a gaseous phase rest all other elements form slag. Slag formation of desired chemistry and physico-chemical properties is vital for the successful operation of converter steelmaking technology.
- Typically converter steelmaking technology allows to tap liquid steel in approximately every 50 to 60 minutes with specified steel chemistry and 500-1000ppm dissolved oxygen.
- Typically oxygen blowing time is independent of converter capacity i.e. O_2 is blown for 15 to 20 minutes irrespective of the converter capacity.

b) Electric Arc furnace (E A F)

In electric arc furnace steelmaking scrap + hot metal + directly reduced iron is used to produce plain carbon steel

Electric energy is the principle source of thermal energy. Graphite electrodes are used to supply the current. The AC electric arc furnaces are very popular. EAF can be either normal power or ultra high power (UHP) with single or twin shell, with or without bottom stirring or post combustion. EAF generates a considerable noise. Now a days EAF has occupied a unique position in the steel industry: EAF can be switched over easily to produce plain C or alloy steel depending on the market requirements.

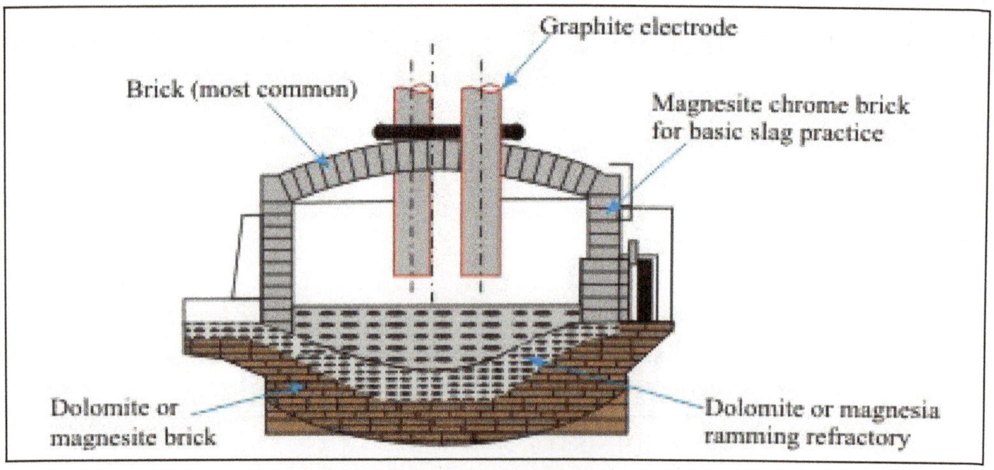

Electric arc furnace

Secondary Steelmaking

The objective of secondary steelmaking is to make the steel of desired chemistry and cleanliness by performing the following treatments in "Ladle":

a) To stir the molten steel by purging inert gas through the bottom of the ladle.

b) To inject slag forming powder either through a lance for further refining

c) To produce clean steel either by removing inclusions or modify them by suitable injecting materials

d) To carry-out deoxidation and degassing.

Secondary steelmaking in ladles has become an integral part of steelmaking. Ladles have additional heating facility and are called Ladle furnaces (LF).

There are several practices adopted for degassing, like vacuum tank degasser, stream degassing and recirculation degassing. In recirculation degassing steel is made to flow from the ladle into a separate degassing chamber and then returned after exposure to the vacuum. In one of the recirculation degassing practice metal circulation is achieved by dipping the degassing vessel into the ladle, the liquid steel is raised into the vessel, degassed and returned into the ladle.

In another practice a refractory lined vessel is equipped with two legs (called snorkels) for dipping into the ladle containing molten steel. Pressure is reduced and argon gas is passed into one of the snorkel, thereby molten steel is raised into the vessel and recirculates back into the ladle through the other snorkel.

Types of Steel

Weathering Steel

weathering steel – Fulcrum (1987) by Richard Serra

Weathering steel, often referred to by the genericized trademark COR-TEN steel and sometimes written without the hyphen as corten steel, is a group of steel alloys which were developed to eliminate the need for painting, and form a stable rust-like appearance if exposed to the weather for several years.

U.S. Steel holds the registered trademark on the name COR-TEN. Although USS sold its discrete plate business to International Steel Group (now Arcelor-Mittal) in 2003, it still sells COR-TEN branded material in strip-mill plate and sheet forms.

The original COR-TEN received the standard designation A 242 (COR-TEN A) from the ASTM International standards group. Newer ASTM grades are A 588 (COR-TEN B) and A 606 for thin sheet. All alloys are in common production and use.

Properties

Weathering refers to the chemical composition of these steels, allowing them to exhibit increased resistance to atmospheric corrosion compared to other steels. This is because the steel forms a protective layer on its surface under the influence of the weather.

The corrosion-retarding effect of the protective layer is produced by the particular distribution and concentration of alloying elements in it. The layer protecting the surface develops and regenerates continuously when subjected to the influence of the weather. In other words, the steel is allowed to rust in order to form the protective coating.

Chemical composition of Cor-ten grades (%, besides iron)									
Grade	**C**	**Si**	**Mn**	**P**	**S**	**Cr**	**Cu**	**V**	**Ni**
Cor-ten A	0.12	0.25–0.75	0.20–0.50	0.01–0.20	0.030	0.50–1.25	0.25–0.55		0.65
Cor-ten B	0.16	0.30–0.50	0.80–1.25	0.030	0.030	0.40–0.65	0.25–0.40	0.02–0.10	0.40

The mechanical properties of weathering steels depend on which alloy and how thick the material is.

ASTM A 242

The original A 242 alloy has a yield strength of 50 kilopounds per square inch (340 MPa) and ultimate tensile strength of 70 ksi (480 MPa) for light-medium rolled shapes and plates up to 0.75 inches (19 mm) thick. It has yield strength of 46 ksi (320 MPa) and ultimate strength of 67 ksi (460 MPa) for medium weight rolled shapes and plates from 0.75–1 inch (19–25 mm) thick. The thickest rolled sections and plates – from 1.5–4 in (38–102 mm) thick have yield strength of 42 ksi (290 MPa) and ultimate strength of 63 ksi (430 MPa).

ASTM A 588

A 588 has a yield strength of at least 50 ksi (340 MPa), and ultimate tensile strength of 70 ksi (480 MPa) for all rolled shapes and plate thicknesses up to 4 in (100 mm) thick. Plates from 4–5 in (102–127 mm) have yield strength at least 46 ksi (320 MPa) and ultimate tensile strength at least 67 ksi (460 MPa), and plates from 5–8 in (127–203 mm) thick have yield strength at least 42 ksi (290 MPa) and ultimate tensile strength at least 63 ksi (430 MPa).

Uses

Weathering steel is popularly used in outdoor sculptures, such as in the large Chicago Picasso sculpture, which stands in the plaza of the Daley Center Courthouse in Chicago, which is also constructed of the same weathering steel and as exterior facades, for its rustic antique appearance. Examples include Barclays Center, Brooklyn, New York, The Angel of the North, Gateshead, UK and the Humanities and Arts complex at Leeds Metropolitan University – Broadcasting Place – Leeds, UK.

Broadcasting Tower, Leeds, United Kingdom

Abetxuko Bridge by J. Sobrino, PEDELTA, Abetxuko, Vitoria, Spain

It is also used in bridge and other large structural applications such as the New River Gorge Bridge, the second span of the Newburgh–Beacon Bridge (1980), and the creation of the Australian Centre for Contemporary Art (ACCA) and MONA.

It is very widely used in marine transportation, in the construction of intermodal containers as well as visible sheet piling along recently widened sections of London's M25 motorway.

The first use of weathering steel for architectural applications was the John Deere World Headquarters in Moline, Illinois. The building was designed by architect Eero Saarinen, and completed in 1964. The main buildings of Odense University, designed by Knud Holscher and Jørgen Vesterholt and built 1971–1976, are clad in weathering steel, earning them the nickname *Rustenborg*. In 1977, Robert Indiana created a Hebrew version of the *Love* sculpture made from weathering steel using the four-letter word ahava for the Israel Museum Art Garden in Jerusalem, Israel. In Denmark, all masts for supporting the catenary on electrified railways are made of weathering steel for aesthetic reasons.

Weathering steel was used in 1971 for the Highliner electric cars built by the St. Louis Car Company for Illinois Central Railroad. The use of weathering steel was seen as a cost-cutting move in comparison with the contemporary railcar standard of stainless steel. A subsequent order in 1979 was built to similar specs, including weathering steel bodies, by Bombardier. The cars were painted, a standard practice for weathering steel railcars. The durability of weathering steel did not live up to expectations, with rust holes appearing in the railcars. Painting may have contributed to the problem, as painted weathering steel is no more corrosion-resistant than conventional steel, because the protective patina will not form in time to prevent corrosion over a localized area of attack such as a small paint failure. These cars have been retired by 2016.

Weathering steel was used to build the exterior of Barclays Center, made up of 12,000 pre-weathered steel panels engineered by ASI Limited & SHoP Construction. The *New York Times* says of the material, "While it can look suspiciously unfinished to the casual observer, it has many fans in the world of art and architecture."

Disadvantages

US Steel Tower

Using weathering steel in construction presents several challenges. Ensuring that weld-points weather at the same rate as the other materials may require special welding techniques or material. Weathering steel is not rustproof in itself. If water is allowed to accumulate in pockets, those areas will experience higher corrosion rates, so provision for drainage must be made. Weathering steel is sensitive to humid subtropical climates. In such environments, it is possible that the protective patina may not stabilize but instead continue to corrode. For example, the former Omni Coliseum, built in 1972 in Atlanta, never stopped rusting, and eventually large holes appeared in the structure. This was a major factor in the decision to demolish it just 25 years after construction. The same thing can happen in environments laden with sea salt. Hawaii's Aloha Stadium, built in 1975, is one example of this. Weathering steel's normal surface weathering can also lead to rust stains on nearby surfaces.

The rate at which some weathering steels form the desired patina varies strongly with the presence of atmospheric pollutants which catalyze corrosion. While the process is generally successful in large urban centers, the weathering rate is much slower in more rural environments. Uris Hall, a social sciences building on Cornell University's main campus in Ithaca, a small town in Upstate New York, did not achieve the predicted surface finish on its Bethlehem Steel Mayari-R weathering steel framing within the predicted time. Rainwater runoff from the slowly rusting steel stained the numerous large windows and increased maintenance costs. Corrosion without the formation of a protective layer apparently led to the need for emergency structural reinforcement and galvanizing in 1974, less than two years after opening.

The U.S. Steel Tower in Pittsburgh, Pennsylvania was constructed by U.S. Steel in part to showcase COR-TEN steel. The initial weathering of the material resulted in a discoloration of the surrounding city sidewalks which is known as "bleeding" or "runoff", as well as other nearby buildings. A cleanup effort was orchestrated by the corporation once weathering was complete to clean the markings. A few of the nearby sidewalks were left uncleaned, and remain a rust color. This problem has been reduced in newer formulations of weathering steel. Staining can be prevented if the structure can be designed so that water does not drain from the steel onto concrete where stains would be visible.

Low-background Steel

A body counting room at the Rocky Flats Plant in Denver, Colorado made entirely from pre-World War II steel.

Low-background steel is any steel produced prior to the detonation of the first atomic bombs in the 1940s and 1950s. With the Trinity test and the Atomic bombing of Hiroshima and Nagasaki in 1945, and then subsequent nuclear weapons testing during the early years of the Cold War, background radiation levels increased across the world. Modern steel is contaminated with radionuclides because its production used atmospheric air. Low background steel is so called because it does not suffer from such nuclear contamination. This steel is used in devices that require the highest sensitivity for detecting radionuclides.

The primary source of low-background steel is ships that have been under water since before the Trinity test, most famously the scuttled German WWI battleships in Scapa Flow.

Radionuclide Contamination

From 1856 until the mid 20th century, steel was produced in the Bessemer process where air was forced into Bessemer converters converting the pig iron into steel. By the mid-20th century, many steelworks had switched to the BOS process which uses pure oxygen instead of air. However as both processes use atmospheric gas, they are susceptible to contamination from airborne particulates. Present-day air carries radionuclides, such as cobalt-60, which are deposited into the steel giving it a weak radioactive signature.

World anthropogenic background radiation levels peaked at 0.15 mSv/yr above natural levels in 1963, the year that the Partial Nuclear Test Ban Treaty was enacted. Since then, anthropogenic background radiation has decreased exponentially to 0.005 mSv/yr above natural levels.

However, cobalt-60 still contaminates modern steel because it has been recycled through the scrap metal supply chain.

Necessity

Devices that require low-background steel include:

- Geiger counters
- Medical apparatus: whole body counting and lung counters
- Scientific equipment: photonics
- Aeronautical and space sensors

As these are detecting radioactive emissions, they require an extremely low radiation environment in order to work correctly. Low background counting chambers are made from low-background steel with extremely heavy radiation shielding. They are used to detect the minutest nuclear emissions.

Structural Steel

Structural steel is a category of steel used as a construction material for making structural steel shapes. A structural steel shape is a profile, formed with a specific cross section and following certain standards for chemical composition and mechanical properties. Structural steel shapes, sizes, composition, strengths, storage practices, etc., are regulated by standards in most industrialized countries.

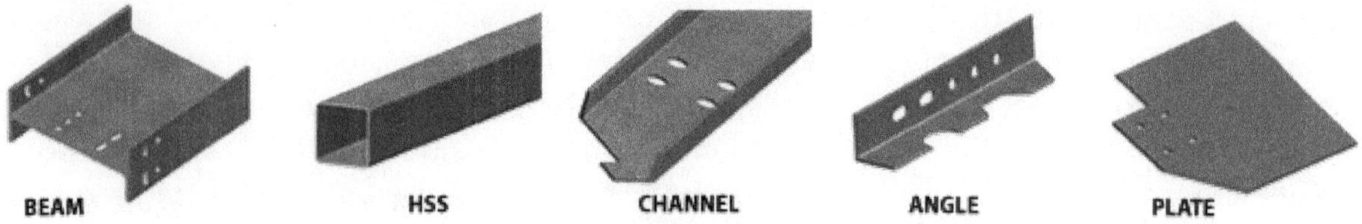

Various structural steel shapes

Structural steel members, such as I-beams, have high second moments of area, which allow them to be very stiff in respect to their cross-sectional area.

Common Structural Shapes

A steel I-beam, in this case used to support timber joists in a house.

- I-beam (I-shaped cross-section - in Britain these include Universal Beams (UB) and Universal Columns (UC); in Europe it includes the IPE, HE, HL, HD and other sections; in the US it includes Wide Flange (WF or W-Shape) and H sections)
- Z-Shape (half a flange in opposite directions)
- HSS-Shape (Hollow structural section also known as SHS (structural hollow section) and including square, rectangular, circular (pipe) and elliptical cross sections)
- Angle (L-shaped cross-section)
- Structural channel, or C-beam, or C cross-section
- Tee (T-shaped cross-section)
- Rail profile (asymmetrical I-beam)
    - Railway rail
    - Vignoles rail
    - Flanged T rail
    - Grooved rail
- Bar, a piece of metal, rectangular cross sectioned (flat) and long, but not so wide so as to be called a sheet.
- Rod, a round or square and long piece of metal.
- Plate, metal sheets thicker than 6 mm or $^1/_4$ in.

- Open web steel joist

While many sections are made by hot or cold rolling, others are made by welding together flat or bent plates (for example, the largest circular hollow sections are made from flat plate bent into a circle and seam-welded).

Standards

Standard Structural Steels (Europe)

Most steels used throughout Europe are specified to comply with the European standard EN 10025. However, many national standards also remain in force.

Typical grades are described as 'S275J2' or 'S355K2W'. In these examples, 'S' denotes structural rather than engineering steel; 275 or 355 denotes the yield strength in newtons per square millimetre or the equivalent megapascals; J2 or K2 denotes the materials toughness by reference to Charpy impact test values; and the 'W' denotes weathering steel. Further letters can be used to designate fine grain steel ('N' or 'NL'); quenched and tempered steel ('Q' or 'QL'); and thermomechanically rolled steel ('M' or 'ML').

1. S275JOH Specification S275JOH is steel grade in EN 10219 specification, EN 10210 standard. And the most widely used specification is EN10219 standard, which is Cold formed welded structural hollow sections of non-alloy and fine grain steels.

EN10219-1 specifies the technical delivery conditions for cold formed welded structural hollow sections of circular, square or rectangular forms and applies to structural hollow sections formed cold without subsequent heat treatment.

Requirements for S275JOH pipe tolerances, dimensions and sectional s275 pipe properties are contained in EN 10219-2.

2. S275JOH Steel Pipes manufacture Process

The steel manufacturing process shall be at the discretion of the steel producer. S275JOH carbon steel pipes can be made in ERW, SAW or seamless process. All S275JOH steel material and S275JOH pipes should conform to EN10219 standards.

The normal yield strength grades available are 195, 235, 275, 355, 420, and 460, although some grades are more commonly used than others e.g. in the UK, almost all structural steel is grades S275 and S355. Higher grades are available in quenched and tempered material (500, 550, 620, 690, 890 and 960 - although grades above 690 receive little if any use in construction at present).

A set of euronorms define the shape of a set of standard structural profiles:

- European I-beam: IPE - Euronorm 19-57
- European I-beam: IPN - DIN 1025-1
- European flange beams: HE - Euronorm 53-62

- European channels: UPN - DIN 1026-1
- European cold formed IS IS 800-1

Standard Structural Steels (USA)

Steels used for building construction in the US use standard alloys identified and specified by ASTM International. These steels have an alloy identification beginning with A and then two, three, or four numbers. The four-number AISI steel grades commonly used for mechanical engineering, machines, and vehicles are a completely different specification series.

The standard commonly used structural steels are:

Carbon Steels

- A36 - structural shapes and plate.
- A53 - structural pipe and tubing.
- A500 - structural pipe and tubing.
- A501 - structural pipe and tubing.
- A529 - structural shapes and plate.

High Strength Low Alloy Steels

- A441 - structural shapes and plates.
- A572 - structural shapes and plates.
- A618 - structural pipe and tubing.
- A992 - Possible applications are W or S I-Beams.
- A913 - Quenched and Self Tempered (QST) W shapes.
- A270 - structural shapes and plates.
- A521 - structural plates and shapes∞

Corrosion Resistant High Strength Low Alloy Steels

- A243 - structural shapes and plates.
- A588 - structural shapes and plates.

Quenched and Tempered Alloy Steels

- A514 - structural shapes and plates.
- A517 - boilers and pressure vessels.
- Eglin steel - Inexpensive aerospace and weaponry items.

Non-preload bolt assembly (EN 15048)

Pre-load bolt assembly (EN 14399)

CE Marking

The concept of CE marking for all construction products and steel products is introduced by the Construction Products Directive (CPD). The CPD is a European Directive that ensures the free movement of all construction products within the European Union.

Because steel components are "safety critical", CE Marking is not allowed unless the Factory Production Control (FPC) system under which they are produced has been assessed by a suitable certification body that has been approved to the European Commission.

In the case of steel products such as sections, bolts and fabricated steelwork the CE Marking demonstrates that the product complies with the relevant harmonized standard.

For steel structures the main harmonized standards are:

- Steel sections and plate - EN 10025-1
- Hollow sections - EN 10219-1 and EN 10210-1
- Pre-loadable bolts - EN 14399-1
- Non-preloadable bolts - EN 15048-1
- Fabricated steel - EN 1090 -1

The standard that covers CE Marking of structural steelwork is EN 1090-1. The standard has come into force in late 2010. After a transition period of two years, CE Marking will become mandatory in most European Countries sometime early in 2012. The official end date of the transition period is July 1, 2014.

Steel vs. Concrete

Choosing the Ideal Structural Material

Most construction projects require the use of hundreds of different materials. These range from concrete of all different specifications, structural steel of different specifications, clay, mortar, ceramics, wood, etc. In terms of a load bearing structural frame, they will generally consist of structural steel, concrete, masonry, and/or wood, using a suitable combination of each to produce an efficient structure. Most commercial and industrial structures are primarily constructed using either structural steel or reinforced concrete. When designing a structure, an engineer must decide which, if not both, material is most suitable for the design. There are many factors considered when choosing a construction material. Cost is commonly the controlling element; however, other considerations such as weight, strength, constructability, availability, sustainability, and fire resistance will be taken into account before a final decision is made.

- Cost - The cost of these construction materials will depend entirely on the geographical location of the project and the availability of the materials. Just as the price of gasoline fluctuates, so do the prices of cement, aggregate, steel, etc. Reinforced concrete derives about half of its construction costs from the required form-work. This refers to the lumber necessary to build the "box" or container in which the concrete is poured and held until it cures. The expense of the forms makes precast concrete a popular option for designers due to the reduced costs and time. Due to the fact that steel is sold by the pound it is the responsibility of the structural designer to specify the lightest members possible while still maintaining a safe structural design. An additional method of reducing expenditures in design is to use many of the same size steel members as opposed to many unique members.

- Strength/weight ratio - Construction materials are commonly categorized by their strength to weight ratio, or specific strength. This is defined as the strength of a material over its density. This gives an engineer an indication as to how useful the material is in comparison to its weight, with the weight being a direct indication of its cost (typically) and ease of construction. Concrete is typically ten times stronger in compression than in tension, giving it a higher strength to weight ratio in compression, only.

- Sustainability - Many construction companies and material vendors are making changes to be a more environmentally friendly company. Sustainability has become an entirely new consideration for materials that are to be placed into the environment for generations of time. A sustainable material will be one that has minimal effect on the environment, both at the time of installation as well as throughout the life cycle of the material. Reinforced concrete and structural steel both have the ability to be a sustainable construction option, if used properly. Over 80% of structural steel members fabricated today come from recycled metals, called A992 steel. This member material is cheaper, as well as having a higher strength to weight ratio, than previously used steel members (A36 grade). Reinforced concrete can also be a very sustainable option. Concrete's material components are naturally occurring materials that are not harmful to the environment. Also, concrete can now be poured to be permeable, allowing the flow of water through a paved surface and therefore reducing the necessary infrastructural construction that causes further potential disturbance to the environment. Concrete can also be crushed and used as aggregate in future concrete applications, meaning it doesn't necessarily have to be used as a land fill.

- Fire resistance - One of the most dangerous hazards to a building is a fire hazard. This is especially true in dry, windy climates and for structures constructed using wood. Special considerations must be taken into account with structural steel to ensure it is not under a dangerous fire hazard condition. Reinforced concrete characteristically does not pose a threat in the event of fire and even resists the spreading of fire, as well as temperature changes. This makes concrete an excellent insulation, improving the sustainability of the building it surrounds by reducing the required energy to maintain climate.
- Corrosion - When choosing a structural material, it is important to consider the life cycle of the building. Some materials are susceptible to corrosion from their surrounding elements, such as water, heat, humidity, or salt. Special considerations must be taken into account during the installation of a structural material to prevent any potential corrosion hazards. This must also be made clear to the occupants of the building because there may or may not be a necessary maintenance requirement to prevent corrosion. For example, structural steel cannot be exposed to the environment because any moisture, or other contact with water, will cause it to rust. When the steel rusts it compromises the structural integrity of the building and poses a potential danger to the residual or surrounding occupants.

Reinforced Concrete

- Characteristics - Generally consisting of portland cement, water, construction aggregate (coarse and fine), and steel reinforcing bars (rebar), concrete is cheaper in comparison to structural steel.
- Strength - Concrete is a composite material with relatively high compressive strength properties, but lacking in tensile strength/ductility. This inherently makes concrete a useful material for carrying the weight of a structure. Concrete reinforced with steel rebar give the structure a stronger tensile capacity, as well as an increase in ductility and elasticity.
- Constructability - Reinforced concrete must be poured and left to set, or harden. After setting (typically 1–2 days), a concrete must cure, the process in which concrete experiences a chemical reaction between the cementitious particles and the water. The curing process is complete after 28 days; however, construction may continue after 1–2 weeks, depending on the nature of the structure. Concrete can be constructed into nearly any shape and size. Approximately half of the cost of using reinforced concrete in a structural project is attributed to the construction of the form-work. In order to save time, and therefore costs, structural concrete members may be pre-cast. This refers to a reinforced concrete beam, girder, or column being poured off site and left to cure. After the curing process, the concrete member may be delivered to the construction site and installed as soon as it is needed. Since the concrete member was cured off location beforehand, construction may continue immediately after erection.

- Fire resistance - Concrete has excellent fire resistance properties, requiring no additional construction costs to adhere to the International Building Code (IBC) fire protection standards. However, concrete buildings will still likely use other materials that are not fire resistant. Therefore, a designer must still take into account the use of the concrete and where it will require fire hazardous materials in order to prevent future complications in the overall design.
- Corrosion - Reinforced concrete, when constructed properly, has excellent corrosion resistance properties. Concrete is not only resistant to water, but needs it to cure and develop its strength over time. However, the steel reinforcement in the concrete must not be exposed in order to prevent its corrosion as this could significantly reduce the ultimate strength of the structure. The American Concrete Institute provides the necessary design specifications for an engineer to ensure there is enough concrete covering any steel reinforcement to prevent exposure to water. This cover distance must be specified because concrete will inevitable crack at locations carrying tension, or locations containing reinforcing bars for the purpose of carrying said tension. If the concrete cracks, it provides a path for water to travel directly to the reinforcing bars. Some reinforcing bars are coated in epoxy as a second order measure of preventing corrosion due to water contact. This method induces higher costs on the overall project, however, due to the higher cost of the epoxy coated bars. Also, when using epoxy coated bars, reinforced concrete members must be designed larger, as well as stronger, in order to balance the loss of friction between the reinforcing bars and concrete. This friction is referred to as bond strength, and it is vital to the structural integrity of a concrete member.

Structural Steel

- Characteristics - Structural steel differs from concrete in its attributed compressive strength as well as tensile strength.
- Strength - Having high strength, stiffness, toughness, and ductile properties, structural steel is one of the most commonly used materials in commercial and industrial building construction.
- Constructability - Structural steel can be developed into nearly any shape, which are either bolted or welded together in construction. Structural steel can be erected as soon as the materials are delivered on site, whereas concrete must be cured at least 1–2 weeks after pouring before construction can continue, making steel a schedule-friendly construction material.
- Fire resistance - Steel is inherently a noncombustible material. However,when heated to temperatures seen in a fire scenario, the strength and stiffness of the material is significantly reduced. The International Building Code requires steel be enveloped in sufficient fire-resistant materials, increasing overall cost of steel structure buildings.
- Corrosion - Steel, when in contact with water, can corrode, creating a potentially dangerous structure. Measures must be taken in structural steel construction to prevent any lifetime corrosion. The steel can be painted, providing water resistance. Also, the fire resistance material used to envelope steel is commonly water resistant.

- Mold - Steel provides a less suitable surface environment for mold to grow than wood.

The tallest structures today (commonly called "skyscrapers" or high-rise) are constructed using structural steel due to its constructability, as well as its high strength-to-weight ratio. In comparison, concrete, while being less dense than steel, has a much lower strength-to-weight ratio. This is due to the much larger volume required for a structural concrete member to support the same load; steel, though denser, does not require as much material to carry a load. However, this advantage becomes insignificant for low-rise buildings, or those with several stories or less. Low-rise buildings distribute much smaller loads than high-rise structures, making concrete the economical choice. This is especially true for simple structures, such as parking garages, or any building that is a simple, rectilinear shape.

Structural steel and reinforced concrete are not always chosen solely because they are the most ideal material for the structure. Companies rely on the ability to turn a profit for any construction project, as do the designers. The price of raw materials (steel, cement, coarse aggregate, fine aggregate, lumber for form-work, etc.) is constantly changing. If a structure could be constructed using either material, the cheapest of the two will likely control. Another significant variable is the location of the project. The closest steel fabrication facility may be much further from the construction site than the nearest concrete supplier. The high cost of energy and transportation will control the selection of the material as well. All of these costs will be taken into consideration before the conceptual design of a construction project is begun.

Combining Steel and Reinforced Concrete

Structures consisting of both materials utilize the benefits of structural steel and reinforced concrete. This is already common practice in reinforced concrete in that the steel reinforcement is used to provide steel's tensile strength capacity to a structural concrete member. A commonly seen example would be parking garages. Some parking garages are constructed using structural steel columns and reinforced concrete slabs. The concrete will be poured for the foundational footings, giving the parking garage a surface to be built on. The steel columns will be connected to the slab by bolting and/or welding them to steel studs extruding from the surface of the poured concrete slab. Pre-cast concrete beams may be delivered on site to be installed for the second floor, after which a concrete slab may be poured for the pavement area. This can be done for multiple stories. A parking garage of this type is just one possible example of many structures that may use both reinforced concrete and structural steel.

A structural engineer understands that there are an infinite number of designs that will produce an efficient, safe, and affordable building. It is the engineer's job to work alongside the owner(s), contractor(s), and all other parties involved to produce an ideal product that suits everyone's needs. When choosing the structural materials for their structure, the engineer has many variables to consider, such as the cost, strength/weight ratio, sustainability of the material, constructability, etc.

Thermal Properties

The properties of steel vary widely, depending on its alloying elements.

The austenizing temperature, the temperature where a steel transforms to an austenite crystal structure, for steel starts at 900 °C (1,650 °F) for pure iron, then, as more carbon is added, the temperature falls to a minimum 724 °C (1,335 °F) for eutectic steel (steel with only .83% by weight of carbon in it). As 2.1% carbon (by mass) is approached, the austenizing temperature climbs back up, to 1,130 °C (2,070 °F). Similarly, the melting point of steel changes based on the alloy.

The lowest temperature at which a plain carbon steel can begin to melt, its solidus, is 1,130 °C (2,070 °F). Steel never turns into a liquid below this temperature. Pure Iron ('Steel' with 0% Carbon) starts to melt at 1,492 °C (2,718 °F), and is completely liquid upon reaching 1,539 °C (2,802 °F). Steel with 2.1% Carbon by weight begins melting at 1,130 °C (2,070 °F), and is completely molten upon reaching 1,315 °C (2,399 °F). 'Steel' with more than 2.1% Carbon is no longer Steel, but is known as Cast iron.

Fire Resistance

Metal deck and open web steel joist receiving spray fireproofing plaster, made of polystyrene-leavened gypsum.

Steel loses strength when heated sufficiently. The *critical temperature* of a steel member is the temperature at which it cannot safely support its load. Building codes and structural engineering standard practice defines different critical temperatures depending on the structural element type, configuration, orientation, and loading characteristics. The critical temperature is often considered the temperature at which its yield stress has been reduced to 60% of the room temperature yield stress. In order to determine the fire resistance rating of a steel member, accepted calculations practice can be used, or a fire test can be performed, the critical temperature of which is set by the standard accepted to the Authority Having Jurisdiction, such as a building code. In Japan, this is below 400 °C. In China, Europe and North America (e.g., ASTM E-119), this is approximately 1000–1300 °F (530-810 °C). The time it takes for the steel element that is being tested to reach the temperature set by the test standard determines the duration of the fire-resistance rating. Heat transfer to the steel can be slowed by the use of fireproofing materials, thus limiting steel temperature. Common fireproofing methods for structural steel include intumescent, endothermic, and plaster coatings as well as drywall, calcium silicate cladding, and mineral wool insulating blankets.

Concrete building structures often meet code required fire-resistance ratings, as the concrete thickness over the steel rebar provides sufficient fire resistance. However, concrete can be subject to spalling, particularly if it has an elevated moisture content. Although additional fireproofing is not often applied to concrete building structures, it is sometimes used in traffic tunnels and locations where a hydrocarbon fuel fire is more likely, as flammable liquid fires provides more heat to the structural element as compared to a fire involving ordinary combustibles during the same fire period. Structural steel fireproofing materials include intumescent, endothermic and plaster coatings as well as drywall, calcium silicate cladding, and mineral or high temperature insulation wool blankets. Attention is given to connections, as the thermal expansion of structural elements can compromise fire-resistance rated assemblies.

Manufacturing

Cutting workpieces to length is usually done with a bandsaw.

A beam drill line (drill line) has long been considered an indispensable way to drill holes and mill slots into beams, channels and HSS elements. CNC beam drill lines are typically equipped with feed conveyors and position sensors to move the element into position for drilling, plus probing capability to determine the precise location where the hole or slot is to be cut.

For cutting irregular openings or non-uniform ends on dimensional (non-plate) elements, a cutting torch is typically used. Oxy-fuel torches are the most common technology and range from simple hand-held torches to automated CNC coping machines that move the torch head around the structural element in accordance with cutting instructions programmed into the machine.

Fabricating flat plate is performed on a plate processing center where the plate is laid flat on a stationary 'table' and different cutting heads traverse the plate from a gantry-style arm or "bridge." The cutting heads can include a punch, drill or torch.

Carbon Steel

Carbon steels are steels with carbon content up to 2.1% by weight. American Iron and Steel Institute (AISI) definition of Carbon Steel states:

> Steel is considered to be carbon steel when:

- no minimum content is specified or required for chromium, cobalt, molybdenum, nickel, niobium, titanium, tungsten, vanadium or zirconium, or any other element to be added to obtain a desired alloying effect;
- the specified minimum for copper does not exceed 0.40 percent;
- or the maximum content specified for any of the following elements does not exceed the percentages noted: manganese 1.65, silicon 0.60, copper 0.60.

The term "carbon steel" may also be used in reference to steel which is not stainless steel; in this use carbon steel may include alloy steels.

As the carbon percentage content rises, steel has the ability to become harder and stronger through heat treating; however, it becomes less ductile. Regardless of the heat treatment, a higher carbon content reduces weldability. In carbon steels, the higher carbon content lowers the melting point.

Type

Mild or Low-carbon Steel

Mild steel (steel containing a small percentage of carbon, strong and tough but not readily tempered), also known as plain-carbon steel and Low carbon steel. It is now the most common form of steel because its price is relatively low while it provides material properties that are acceptable for many applications. Mild steel contains approximately 0.05–0.25% carbon making it malleable and ductile. Mild steel has a relatively low tensile strength, but it is cheap and easy to form; surface hardness can be increased through carburizing.

It is often used when large quantities of steel are needed, for example as structural steel. The density of mild steel is approximately 7.85 g/cm^3 (7850 kg/m^3 or 0.284 lb/in^3) and the Young's modulus is 200 GPa (29,000,000 psi).

Low-carbon steels suffer from *yield-point runout* where the material has two yield points. The first yield point (or upper yield point) is higher than the second and the yield drops dramatically after the upper yield point. If a low-carbon steel is only stressed to some point between the upper and lower yield point then the surface develop Lüder bands. Low-carbon steels contain less carbon than other steels and are easier to cold-form, making them easier to handle.

Higher-carbon Steels

Carbon steels which can successfully undergo heat-treatment have a carbon content in the range of 0.30–1.70% by weight. Trace impurities of various other elements can have a significant effect on the quality of the resulting steel. Trace amounts of sulfur in particular make the steel red-short, that is, brittle and crumbly at working temperatures. Low-alloy carbon steel, such as A36 grade, contains about 0.05% sulfur and melts around 1,426–1,538 °C (2,599–2,800 °F). Manganese is often added to improve the hardenability of low-carbon steels. These additions turn the material into a low-alloy steel by some definitions, but AISI's definition of carbon steel allows up to 1.65% manganese by weight.

AISI Classification

Carbon steel is broken down into four classes based on carbon content:

Low-carbon Steel

0.05 to 0.25% carbon content.

Medium-carbon Steel

Approximately 0.3–0.6% carbon content. Balances ductility and strength and has good wear resistance; used for large parts, forging and automotive components.

High-carbon Steel

Approximately 0.6–1.0% carbon content. Very strong, used for springs, swords, and high-strength wires.

Ultra-high-carbon Steel

Approximately 1.25–2.0% carbon content. Steels that can be tempered to great hardness. Used for special purposes like (non-industrial-purpose) knives, axles or punches. Most steels with more than 2.5% carbon content are made using powder metallurgy.

Heat Treatment

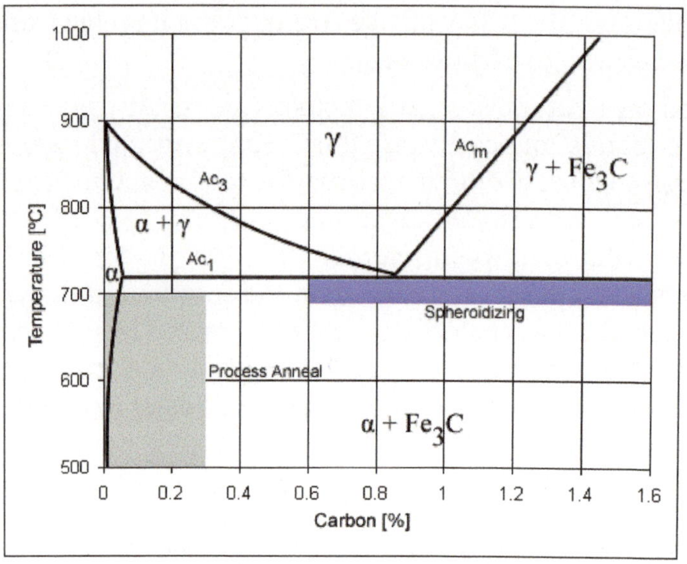

Iron-carbon phase diagram, showing the temperature and carbon ranges for certain types of heat treatments.

The purpose of heat treating carbon steel is to change the mechanical properties of steel, usually ductility, hardness, yield strength, or impact resistance. Note that the electrical and thermal conductivity are only slightly altered. As with most strengthening techniques for steel, Young's modulus (elasticity) is unaffected. All treatments of steel trade ductility for increased strength and vice versa. Iron has a higher solubility for carbon in the austenite phase; therefore all heat treatments, except spheroidizing and process annealing, start by heating the steel to a temperature at which the austenitic phase can exist. The steel is then quenched (heat drawn out) at a moderate to low rate allowing carbon to diffuse out of the austenite forming iron-carbide (cementite) to precipitate leaving ferrite, or at a high rate, trapping the carbon within the iron thus forming martensite. The rate at which the steel is cooled through the eutectoid temperature (about 727°C) affects the rate at which carbon diffuses out of austenite and forms cementite. Generally speaking, cooling swiftly will leave iron carbide finely dispersed and produce a fine grained pearlite and cooling slowly will give a coarser pearlite. Cooling a hypoeutectoid steel (less than 0.77 wt% C) results in a lamellar-pearlitic structure of iron carbide layers with α-ferrite (nearly pure iron) between. If it is hypereutectoid steel (more than 0.77 wt% C) then the structure is full pearlite with small grains (larger than the pearlite lamella) of cementite formed on the grain boundaries. A euctoid steel (0.77% carbon) will have a pearlite structure throughout the grains with no cementite at the boundaries. The relative amounts of constituents are found using the lever rule. The following is a list of the types of heat treatments possible:

- Spheroidizing: Spheroidite forms when carbon steel is heated to approximately 700 °C for over 30 hours. Spheroidite can form at lower temperatures but the time needed drastically increases, as this is a diffusion-controlled process. The result is a structure of rods or spheres of cementite within primary structure (ferrite or pearlite, depending on which side of the eutectoid you are on). The purpose is to soften higher carbon steels and allow more formability. This is the softest and most ductile form of steel.

- Full annealing: Carbon steel is heated to approximately 40 °C above Ac3? or Acm? for 1 hour; this ensures all the ferrite transforms into austenite (although cementite might still exist if the carbon content is greater than the eutectoid). The steel must then be cooled slowly, in the realm of 20 °C (36 °F) per hour. Usually it is just furnace cooled, where the furnace is turned off with the steel still inside. This results in a coarse pearlitic structure, which means the "bands" of pearlite are thick. Fully annealed steel is soft and ductile, with no internal stresses, which is often necessary for cost-effective forming. Only spheroidized steel is softer and more ductile.
- Process annealing: A process used to relieve stress in a cold-worked carbon steel with less than 0.3% C. The steel is usually heated to 550–650 °C for 1 hour, but sometimes temperatures as high as 700 °C.
- Isothermal annealing: It is a process in which hypoeutectoid steel is heated above the upper critical temperature. This temperature is maintained for a time and then reduced to below the lower critical temperature and is again maintained. It is then cooled to room temperature. This method eliminates any temperature gradient.
- Normalizing: Carbon steel is heated to approximately 55 °C above Ac3 or Acm for 1 hour; this ensures the steel completely transforms to austenite. The steel is then air-cooled, which is a cooling rate of approximately 38 °C (100 °F) per minute. This results in a fine pearlitic structure, and a more-uniform structure. Normalized steel has a higher strength than annealed steel; it has a relatively high strength and hardness.
- Quenching: Carbon steel with at least 0.4 wt% C is heated to normalizing temperatures and then rapidly cooled (quenched) in water, brine, or oil to the critical temperature. The critical temperature is dependent on the carbon content, but as a general rule is lower as the carbon content increases. This results in a martensitic structure; a form of steel that possesses a super-saturated carbon content in a deformed body-centered cubic (BCC) crystalline structure, properly termed body-centered tetragonal (BCT), with much internal stress. Thus quenched steel is extremely hard but brittle, usually too brittle for practical purposes. These internal stresses mat cause stress cracks on the surface. Quenched steel is approximately three to four (with more carbon) fold harder than normalized steel.
- Martempering (Marquenching): Martempering is not actually a tempering procedure, hence the term "marquenching". It is a form of isothermal heat treatment applied after an initial quench, typically in a molten salt bath, at a temperature just above the "martensite start temperature". At this temperature, residual stresses within the material are relieved and some bainite may be formed from the retained austenite which did not have time to transform into anything else. In industry, this is a process used to control the ductility and

hardness of a material. With longer marquenching, the ductility increases with a minimal loss in strength; the steel is held in this solution until the inner and outer temperatures of the part equalize. Then the steel is cooled at a moderate speed to keep the temperature gradient minimal. Not only does this process reduce internal stresses and stress cracks, but it also increases the impact resistance.

- Tempering: This is the most common heat treatment encountered, because the final properties can be precisely determined by the temperature and time of the tempering. Tempering involves reheating quenched steel to a temperature below the eutectoid temperature then cooling. The elevated temperature allows very small amounts of spheroidite to form, which restores ductility, but reduces hardness. Actual temperatures and times are carefully chosen for each composition.

- Austempering: The austempering process is the same as martempering, except the quench is interrupted and the steel is held in the molten salt bath at temperatures between 205°C and 540°C, and then cooled at a moderate rate. The resulting steel, called bainite, produces an acicular microstructure in the steel that has great strength (but less than martinsite), greater ductility, higher impact resistance, and less distortion than martinsite steel. The disadvantage of austempering is it can only be used on a few steels, and it requires a special salt bath.

Case Hardening

Case hardening processes harden only the exterior of the steel part, creating a hard, wear resistant skin (the "case") but preserving a tough and ductile interior. Carbon steels are not very hardenable meaning they can not be hardened throughout thick sections. Alloy steels have a better hardenability, so they can through-hardened and do not require case hardening. This property of carbon steel can be beneficial, because it gives the surface good wear characteristics but leaves the core tough.

Forging Temperature of Steel

Steel Type	Maximum forging temperature (°F / °C)	Burning temperature (°F / °C)
1.5% carbon	1920 / 1049	2080 / 1140
1.1% carbon	1980 / 1082	2140 / 1171
0.9% carbon	2050 / 1121	2230 / 1221
0.5% carbon	2280 / 1249	2460 / 1349
0.2% carbon	2410 / 1321	2680 / 1471
3.0% nickel steel	2280 / 1249	2500 / 1371
3.0% nickel–chromium steel	2280 / 1249	2500 / 1371
5.0% nickel (case-hardening) steel	2320 / 1271	2640 / 1449
Chromium–vanadium steel	2280 / 1249	2460 / 1349
High-speed steel	2370 / 1299	2520 / 1385

Stainless steel	2340 / 1282	2520 / 1385
Austenitic chromium–nickel steel	2370 / 1299	2590 / 1420
Silico-manganese spring steel	2280 / 1249	2460 / 1350

Alloy Steel

Alloy steel is steel that is alloyed with a variety of elements in total amounts between 1.0% and 50% by weight to improve its mechanical properties. Alloy steels are broken down into two groups: low-alloy steels and high-alloy steels. The difference between the two is somewhat arbitrary: Smith and Hashemi define the difference at 4.0%, while Degarmo, *et al.*, define it at 8.0%. Most commonly, the phrase "alloy steel" refers to low-alloy steels.

Strictly speaking, every steel is an alloy, but not all steels are called "alloy steels". The simplest steels are iron (Fe) alloyed with carbon (C) (about 0.1% to 1%, depending on type). However, the term "alloy steel" is the standard term referring to steels with *other* alloying elements added deliberately *in addition to* the carbon. Common alloyants include manganese (the most common one), nickel, chromium, molybdenum, vanadium, silicon, and boron. Less common alloyants include aluminum, cobalt, copper, cerium, niobium, titanium, tungsten, tin, zinc, lead, and zirconium.

The following is a range of improved properties in alloy steels (as compared to carbon steels): strength, hardness, toughness, wear resistance, corrosion resistance, hardenability, and hot hardness. To achieve some of these improved properties the metal may require heat treating.

Some of these find uses in exotic and highly-demanding applications, such as in the turbine blades of jet engines, in spacecraft, and in nuclear reactors. Because of the ferromagnetic properties of iron, some steel alloys find important applications where their responses to magnetism are very important, including in electric motors and in transformers.

Low-alloy Steels

A few common low alloy steels are:

- D6AC
- 300M
- 256A

Principal Low-alloy Steels	
SAE designation	**Composition**
13xx	Mn 1.75%
40xx	Mo 0.20% or 0.25% or 0.25% Mo & 0.042% S
41xx	Cr 0.50% or 0.80% or 0.95%, Mo 0.12% or 0.20% or 0.25% or 0.30%

43xx	Ni 1.82%, Cr 0.50% to 0.80%, Mo 0.25%
44xx	Mo 0.40% or 0.52%
46xx	Ni 0.85% or 1.82%, Mo 0.20% or 0.25%
47xx	Ni 1.05%, Cr 0.45%, Mo 0.20% or 0.35%
48xx	Ni 3.50%, Mo 0.25%
50xx	Cr 0.27% or 0.40% or 0.50% or 0.65%
50xxx	Cr 0.50%, C 1.00% min
50Bxx	Cr 0.28% or 0.50%, and added boron
51xx	Cr 0.80% or 0.87% or 0.92% or 1.00% or 1.05%
51xxx	Cr 1.02%, C 1.00% min
51Bxx	Cr 0.80%, and added boron
52xxx	Cr 1.45%, C 1.00% min
61xx	Cr 0.60% or 0.80% or 0.95%, V 0.10% or 0.15% min
86xx	Ni 0.55%, Cr 0.50%, Mo 0.20%
87xx	Ni 0.55%, Cr 0.50%, Mo 0.25%
88xx	Ni 0.55%, Cr 0.50%, Mo 0.35%
92xx	Si 1.40% or 2.00%, Mn 0.65% or 0.82% or 0.85%, Cr 0.00% or 0.65%
94Bxx	Ni 0.45%, Cr 0.40%, Mo 0.12%, and added boron
ES-1	Ni 5%, Cr 2%, Si 1.25%, W 1%, Mn 0.85%, Mo 0.55%, Cu 0.5%, Cr 0.40%, C 0.2%, V 0.1%

Material Science

Alloying elements are added to achieve certain properties in the material. As a guideline, alloying elements are added in lower percentages (less than 5%) to increase strength or hardenability, or in larger percentages (over 5%) to achieve special properties, such as corrosion resistance or extreme temperature stability. Manganese, silicon, or aluminum are added during the steelmaking process to remove dissolved oxygen, sulfur and phosphorus from the melt. Manganese, silicon, nickel, and copper are added to increase strength by forming solid solutions in ferrite. Chromium, vanadium, molybdenum, and tungsten increase strength by forming second-phase carbides. Nickel and copper improve corrosion resistance in small quantities. Molybdenum helps to resist embrittlement. Zirconium, cerium, and calcium increase toughness by controlling the shape of inclusions. Sulfur (in the form of manganese sulfide) lead, bismuth, selenium, and tellurium increase machinability. The alloying elements tend to form either solid solutions or compounds or carbides. Nickel is very soluble in ferrite; therefore, it forms compounds, usually Ni_3Al. Aluminium dissolves in the ferrite and forms the compounds Al_2O_3 and AlN. Silicon is also very soluble and usually forms the compound $SiO_2 \cdot M_xO_y$. Manganese mostly dissolves in ferrite forming the compounds MnS, $MnO \cdot SiO_2$, but will also form carbides in the form of $(Fe,Mn)_3C$. Chromium forms partitions between the ferrite and carbide phases in steel, forming $(Fe,Cr_3)C$, Cr_7C_3, and $Cr_{23}C_6$. The type of carbide that chromium forms depends on the amount of carbon and other types of alloying elements present. Tungsten and molybdenum form carbides if there is enough carbon and an absence of stronger carbide forming elements (i.e., titanium & niobium), they form the carbides W_2C and Mo_2C, respectively. Vanadium, titanium, and niobium are strong carbide forming elements, forming vanadium carbide, titanium carbide, and niobium carbide, respectively. Alloying elements also have an effect on the eutectoid temperature of the steel. Manganese and nickel lower the eutectoid temperature and are known as *austenite stabilizing elements*. With enough of these elements the austenitic structure may be obtained at room temperature. Carbide-forming elements raise the eutectoid temperature; these elements are known as *ferrite stabilizing elements*.

Principal effects of major alloying elements for steel

Element	Percentage	Primary function
Aluminium	0.95–1.30	Alloying element in nitriding steels
Bismuth	-	Improves machinability
Boron	0.001–0.003	A powerful hardenability agent
Chromium	0.5–2	Increases hardenability
	4–18	Increases corrosion resistance
Copper	0.1–0.4	Corrosion resistance
Lead	-	Improved machinability
Manganese	0.25–0.40	Combines with sulfur and with phosphorus to reduce the brittleness. Also helps to remove excess oxygen from molten steel.
	>1	Increases hardenability by lowering transformation points and causing transformations to be sluggish
Molybdenum	0.2–5	Stable carbides; inhibits grain growth. Increases the toughness of steel, thus making molybdenum a very valuable alloy metal for making the cutting parts of machine tools and also the turbine blades of turbojet engines. Also used in rocket motors.
Nickel	2–5	Toughener
	12–20	Increases corrosion resistance
Silicon	0.2–0.7	Increases strength
	2.0	Spring steels
	Higher percentages	Improves magnetic properties
Sulfur	0.08–0.15	Free-machining properties
Titanium	-	Fixes carbon in inert particles; reduces martensitic hardness in chromium steels
Tungsten	-	Also increases the melting point.
Vanadium	0.15	Stable carbides; increases strength while retaining ductility; promotes fine grain structure. Increases the toughness at high temperatures

Damascus Steel

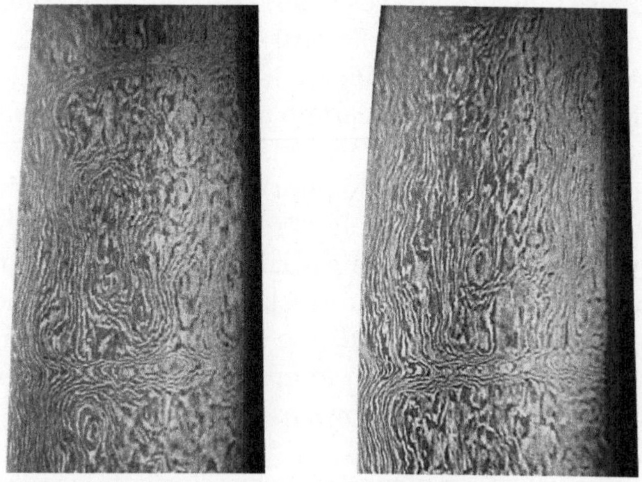

Close-up of an 18th-century Persian-forged Damascus steel sword

Damascus steel was a type of steel used for manufacturing sword blades in the Near East made with wootz steel. These swords are characterized by distinctive patterns of banding and mottling reminiscent of flowing water. Such blades were reputed to be tough, resistant to shattering and capable of being honed to a sharp, resilient edge.

The steel is named after Damascus, the capital city of Syria. It may either refer to swords made or sold in Damascus directly, or it may just refer to the aspect of the typical patterns, by comparison with Damask fabrics (which are in turn named after Damascus).

The original method of producing Damascus steel is not known. Modern attempts to duplicate the metal have not been entirely successful due to differences in raw materials and manufacturing techniques. Several individuals in modern times have claimed that they have rediscovered the methods by which the original Damascus steel was produced.

The reputation and history of Damascus steel has given rise to many legends, such as the ability to cut through a rifle barrel or to cut a hair falling across the blade. A research team in Germany published a report in 2006 revealing nanowires and carbon nanotubes in a blade forged from Damascus steel. Although many types of modern steel outperform ancient Damascus alloys, chemical reactions in the production process made the blades extraordinary for their time, as Damascus steel was superplastic and very hard at the same time. During the smelting process to obtain Wootz steel ingots, woody biomass and leaves are known to have been used as carburizing additives along with certain specific types of iron rich in microalloying elements. These ingots would then be further forged and worked into Damascus steel blades. Research now shows that carbon nanotubes can be derived from plant fibers, suggesting how the nanotubes were formed in the steel. Some experts expect to discover such nanotubes in more relics as they are analyzed more closely.

History

Damascus blades were manufactured in the Near East from ingots of wootz steel that were imported from India, as well as Sri Lanka. Archaeological evidence suggests that the crucible steel process started in the present-day Tamil Nadu before the start of Common Era. The Arabs introduced the Indian wootz steel to Damascus, where a weapons industry thrived. From the 3rd century to the 17th century, India was shipping steel ingots to the Middle East.

A bladesmith from Damascus, ca. 1900

Loss of the Technique

Production of these patterned swords gradually declined, ceasing by around 1750, and the process was lost to metalsmiths. Several modern theories have ventured to explain this decline, including the breakdown of trade routes to supply the needed metals, the lack of trace impurities in the metals, the possible loss of knowledge on the crafting techniques through secrecy and lack of transmission, suppression of the industry in India by the British Raj, or a combination of all the above.

The original Damascus steel or wootz was imported from India to Damascus, where bladesmiths learned how to forge them into swords. Due to the distance of trade for this steel, a sufficiently lengthy disruption of the trade routes could have ended the production of Damascus steel and eventually led to the loss of the technique in India. As well, the need for key trace impurities of tungsten or vanadium within the materials needed for production of the steel may be absent if this material was acquired from different production regions or smelted from ores lacking these key trace elements. The technique for controlled thermal cycling after the initial forging at a specific temperature could also have been lost, thereby preventing the final damask pattern in the steel from occurring.

The discovery of carbon nanotubes in the Damascus steel's composition supports this hypothesis, since the precipitation of carbon nanotubes probably resulted from a specific process that may be difficult to replicate should the production technique or raw materials used be significantly altered.

Reproduction

A bladesmith forging a Damascus blade

Recreating Damascus steel is a subfield of experimental archaeology. Many have attempted to discover or reverse-engineer the process by which it was made.

Moran: Billet Welding

Characteristic “organic” pattern of Damascus steel

Since the well-known technique of pattern welding produced surface patterns similar to those found on Damascus blades, some blacksmiths were erroneously led to believe that Damascus blades were made using this technique, but today, the difference between wootz steel and pattern welding is fully documented and well understood. Pattern-welded steel has been referred to as "Damascus steel" since 1973 when Bladesmith William F. Moran unveiled his "Damascus knives" at the Knifemakers' Guild Show.

This "Modern Damascus" is made from several types of steel and iron slices welded together to form a billet, and currently the term "damascus" (although technically incorrect) is widely accepted to describe modern pattern welded steel blades in the trade. The patterns vary depending on how the smith works the billet. The billet is drawn out and folded until the desired number of layers are formed. In order to attain a Master Smith rating with the American Bladesmith Society that Moran founded, the smith must forge a damascus blade with a minimum of 300 layers.

Verhoeven and Pendray: Crucible

J. D. Verhoeven and A. H. Pendray published an article on their attempts to reproduce the elemental, structural, and visual characteristics of Damascus steel. They started with a cake of steel that matched the properties of the original wootz steel from India, which also matched a number of original Damascus swords to which Verhoeven and Pendray had access. The wootz was in a soft, annealed state, with a grain structure and beads of pure iron carbide which were the result of its hypereutectoid state. Verhoeven and Pendray had already determined that the grains on the surface of the steel were grains of iron carbide—their goal was to reproduce the iron carbide patterns they saw in the Damascus blades from the grains in the wootz.

Although such material could be worked at low temperatures to produce the striated Damascene pattern of intermixed ferrite and cementite bands in a manner identical to pattern-welded Damascus steel, any heat treatment sufficient to dissolve the carbides would permanently destroy the pattern. However, Verhoeven and Pendray discovered that in samples of true Damascus steel, the Damascene pattern could be recovered by aging at a moderate temperature. They found that certain carbide forming elements, one of which was vanadium, did not disperse until the steel reached higher temperatures than those needed to dissolve the carbides. Therefore, a high heat treatment could remove the visual evidence of patterning associated with carbides but did not remove the underlying patterning of the carbide forming elements; a subsequent lower-temperature heat treatment, at a temperature at which the carbides were again stable, could recover the structure by the binding of carbon by those elements.

Anosov, Wadsworth and Sherby: Bulat

In Russia, chronicles record the use of a material known as bulat steel to make highly valued weapons, including swords, knives and axes. Tsar Michael of Russia reportedly had a bulat helmet made for him in 1621. The exact origin or the manufacturing process of bulat is unknown, but it was likely imported to Russia via Persia and Turkestan, and it was similar and possibly the same as damascus steel. Pavel Petrovich Anosov made several attempts to reproduce the process in the mid-19th century. Wadsworth and Sherby also researched the reproduction of Bulat steel and published their results in 1980.

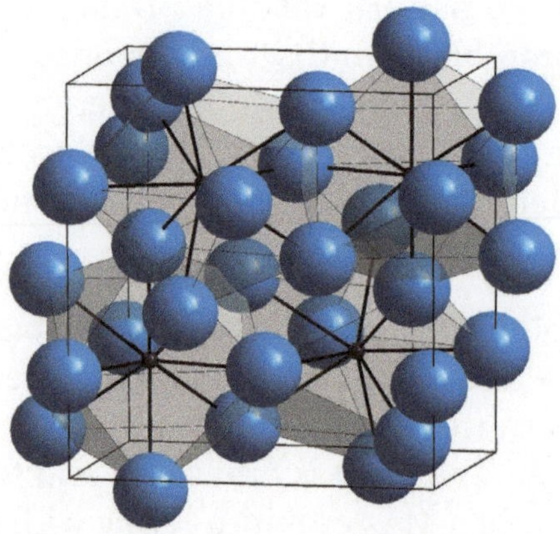

Cementite crystal structure. Iron atoms are in blue, carbon atoms are in black.

Additional Research

A team of researchers based at the Technical University of Dresden that used x-rays and electron microscopy to examine Damascus steel discovered the presence of cementite nanowires and carbon nanotubes. Peter Paufler, a member of the Dresden team, says that these nanostructures are a result of the forging process.

Sanderson proposes that the process of forging and annealing accounts for the nano-scale structures.

In Gunmaking

Prior to the early 20th century, all shotgun barrels were forged by heating narrow strips of iron and steel and shaping them around a mandrel. This process was referred to as "laminating" or "Damascus". These types of barrels earned a reputation for weakness and were never meant to be used with modern smokeless powder, or any kind of moderately powerful explosive. Because of the resemblance to Damascus steel, higher-end barrels were made by Belgian and British gun makers. These barrels are proof marked and meant to be used with light pressure loads. Current gun manufacturers make slide assemblies and small parts such as triggers and safeties for Colt M1911 pistols from powdered Swedish steel resulting in a swirling two-toned effect; these parts are often referred to as "Stainless Damascus".

Science base of Steelmaking

In steelmaking, the impurities like carbon, silicon, manganese, phosphorus and sulphur are removed from hot metal through a combination of gas/metal, gas/slag and gas/metal/slag reactions so as to produce steel of desired chemistry and cleanliness (cleanliness refers to the inclusions). Science of steelmaking involves equilibrium concentration of an impurity between the phases for a given set of temperature and pressure,and the rate of transfer of an impurity from the hot metal.

Equilibrium Between the Phases

The phases in steelmaking are hot metal, molten slag and gas. Hot metal is a multi-component solution in which impurities like carbon, silicon, manganese, phosphorus and sulphur are dissolved in very low amount (total concentration of all the impurities is approximately 5% to 6%) in iron. Slag is a solution of predominantly oxides with small amounts of sulphides, phosphides, silicates etc. Composition of the solutions in steelmaking is conveniently expressed either as weight% (Wt%) or mole fraction(N). The mole fraction of the ith component in a solution of n components is

$$N_i = X_i / \Sigma X_i \qquad (1)$$

where X_i is the number of moles of ith component. The equilibrium of a component between the liquid phases is expressed in terms of integral molar free energy. Integral molar free energy of solution

$$G^m = \Sigma G_i^m N_i = \Sigma G_i N_i - \Sigma G_i^o N_i = RT\Sigma N_i \ln a_i, \qquad (2)$$

$G_i N_i$ represents free energy of solution and $G_i^o N_i$ is the free energy of pure components before entering into the solution. The quantity G_i^m is the partial molar free energy of mixing of component i and represent the change of energy or work which a mole of pure component i can make available.

Chemical potential is a useful concept to describe chemical equilibrium between liquid phases. At chemical equilibrium the chemical potential of any component is identical in all phases. Knowledge of chemical potential is important in steelmaking because an impurity can transfer in the gaseous or slag phase only when its chemical potential is lower than in hot metal.

The criterion for equilibrium at constant temperature and pressure is the change in the integral molar free energy of the solution, $(dG)_{T,P}$ i.e.

$(dG)_{T,P} = 0$ for an infinitesimal process and (3)

$(\Delta G)_{T,P} = 0$ for a finite process

Where $(dG)_{T,P,}$ is change in integral molar free energy

At constant temperature and pressure when $(dG)_{T,P,} < 0$, a process occurs spontaneously. For an isothermal chemical reaction say $A + B = C + D_2 (\Delta G) = \Delta G^- + RTinJ$, where J is activity quotient and ΔG^0 is the standard free energy change. At equilibrium

$\Delta G^0 = -RTIn(J)_{eq} = -RTInK$, where K is equilibrium constant. (4)

Activity of Solution

In dealing with chemical reactions in solution it is important to define the activity of a component. Activity of a component denotes its effective concentration. It is related to fugacity as

$$a_i = f_i / f_i^o \quad (5)$$

f_i is the fugacity of component i in solution and f_i^o is the fugacity of a component in its standard state (standard state could be either pure element or compound at 1 atmospheric pressure) So at standard state activity equals 1. In an ideal gas activity of a component i is equal to its partial pressure.

Raoults's Law

An ideal solution obeys Raoult's law, in which activity of a component a_i equals to its mole fraction N_i

$$a_i = N_i \quad (6)$$

Real solutions exhibit either positive or negative deviation from Raoult's law for a binary solution. Deviation from Raoult' law is taken care by activity coefficient γ_i

$$\gamma_i = a_i / N_i \quad (7)$$

The Fe-Mn forms an ideal solution, whereas the Fe-Cu exhibits strong positive deviation and the Fe-Si strong negative from Raoult's law. Physically it implies that in Fe-Cu solution copper has a strong tendency to segregate, and in Fe-Si solution silicon has a strong tendency to form chemical compound with iron.

In binary liquid oxides, FeO-MnO behaves ideally, whereas most binary silicates i.e. $CaO - SiO_2, FeO - SiO_2, MgO - SiO_2$ show negative deviation from Raoult's law.

Henry's Law

Liquid steel, and to a reasonable extent hot metal primarily fall in the category of dilute solution. In a dilute binary solution activity of a solute obeys Henry's law, which is stated as

$$a_i = \gamma_i^o N_i \quad (8)$$

where γ_i^0 is a constant (activity coefficient for the solute in dilute binary) and N_i is the mole fraction of the specie i. Solutes in all infinite dilute solutions obey Henry's law. Deviation from Henry's law occurs when the solute concentration increases.

In steelmaking the concentration of solute in molten steel is expressed in weight percent. It is frequently most convenient to choose the infinitely dilute solution expressed in terms of weight percent as the standard state. This is defined as

$$h_i / (Wt\%i) = 1 \text{ when } wt\%i \rightarrow 0 \quad (9)$$

For weight percent i other than zero

$$h_i = f_i Wt\%_i \quad (10)$$

Interaction Parameter

Molten steel contains several dissolved solutes in dilute scale. For example, molten steel contains C, S, P, Si, Mn etc. This steel is a multi-component solution. In multi-component solution solutes interact with one another and thus influence activities of other solutes. If Fe is the solvent, and 1, 2....k are solutes in dilute state, then

$$\text{Log } f_i = e_1^1 \text{Wt\%1} + e_1^2 \text{Wt\%2} + e_1^3 \text{Wt\%3} + e_1^4 \text{Wt\%4} + \ldots e_1^J \text{Wt\%J}$$

The term e_i^j is known as interaction parameter describing the influence of solute j on the activity coefficient of solute i. The value of interaction parameter can be found in any book on thermodynamics.

The concept of interaction parameter is very important in estimating the activity of a solute element in presence of other solute elements. For example we want to calculate the activity of sulphur in hot metal of composition C = 4%, Si = 1.5%, Mn = 1% and S = 0.04% at 1600°C. By assuming infinite dilute solution as the standard state, the activity of sulphur is given by

$$h_s = f_s \text{Wt\%S}$$

$$\log f_s = e_s^S \text{Wt\%S} + e_s^C \text{Wt\%C} + e_s^{Si} \text{Wt\%Si} + e_s^{Mn} \text{Wt\%Mn}$$

Substituting the value of $e_s^S = -0.028, e_s^C = 0.24, eS^{Si} = 0.066$ and $eS^{Mn} = -0.025$ we get $f_s = 10.78$ and activity of sulphur is 0.43.

References

- COR-TEN – Weather & Corrosion Resistant Steel: Technical Data, archived from the original on 2010-01-13, retrieved 2010-01-13
- Nishimura, Naoya; Murase, Katsuhiko; Ito, Toshihiro; Watanabe, Takeru; Nowak, Roman. "Ultrasonic detection of spall damage induced by low-velocity repeated impact". Central European Journal of Engineering. 2 (4): 650–655. doi:10.2478/s13531-012-0013-5
- Knowles, Peter Reginald (1987), Design of structural steelwork (2nd ed.), Taylor & Francis, p. 1, ISBN 978-0-903384-59-9
- Olmstead, Elizabeth (2 October 1973). "'Old Rusty' Design Hikes Upkeep Costs". The Cornell Daily Sun. Ithaca, NY. Retrieved 2015-03-29
- J. D. Verhoeven, A. H. Pendray, and W. E. Dauksch; Pendray; Dauksch (1998). "The key role of impurities in ancient damascus steel blades". Journal of Metallurgy. 50 (9): 58. Bibcode:1998JOM....50i..58V. doi:10.1007/s11837-998-0419-y
- Loveless, Robert; Richard Barney (1995) [1977]. How to Make Knives. Knife World Publications. p. 169. ISBN 0-695-80913-X
- Maryon, Herbert (February 1960). "Pattern-Welding and Damascening of Sword-Blades—Part 1: Pattern-Welding". Studies in Conservation. 5 (1): 25–37. doi:10.2307/1505063. JSTOR 1505063
- Brady, George S.; Clauser, Henry R.; Vaccari A., John (1997). Materials Handbook (14th ed.). New York, NY: McGraw-Hill. ISBN 0-07-007084-9
- Sennet, Charles (8 May 1974). "Donor Wrote Corson to Demand 'Old Rusty' Repairs Be Secret". The Cornell Daily Sun. Ithaca, NY. Retrieved 2015-03-29
- Goodell, B, Xie, X., Qian, Y., Daniel, G., Peterson, M., and J. Jellison (2008). "Carbon nanotubes produced from

natural cellulosic materials" (PDF). Journal of Nanoscience and Nanotechnology. 8: 2472–2474. doi:10.1166/jnn.2008.235. PMID 18572666

- Pacey, Arnold (1991). Technology in World Civilization: A Thousand-year History. MIT Press. p. 80. ISBN 978-0-262-66072-3
- Legendary Swords' Sharpness, Strength From Nanotubes, Study Says. National Geographic (2010-10-28). Retrieved on 2011-11-13
- Williams, Alan R. (2003). The knight and the blast furnace: a history of the metallurgy of armour in the Middle Ages & the early modern period Volume 12 of History of warfare. Leiden: BRILL. pp. 11–15. ISBN 978-90-04-12498-1
- Jeffrey Wadsworth; Oleg D. Sherby (1980). "On the Bulat – Damascus Steel Revisited". Prog. Mater. Sci. 25 (1): 35–68. doi:10.1016/0079-6425(80)90014-6
- Sinopoli, Carla M. (2003). The Political Economy of Craft Production: Crafting Empire in South India, c. 1350–1650. Cambridge University Press. p. 192. ISBN 0-521-82613-6
- Kochmann, W.; Reibold, Marianne; Goldberg, Rolf; Hauffe, Wolfgang; Levin, Alexander A; Meyer, Dirk C; Stephan, Thurid; Müller, Heide; Belger, André; Paufler, Peter (2004). "Nanowires in ancient Damascus steel". Journal of Alloys and Compounds. 372: L15–L19. doi:10.1016/j.jallcom.2003.10.005. ISSN 0925-8388
- Juleff, G. (1996). "An ancient wind powered iron smelting technology in Sri Lanka". Nature. 379 (6560): 60. Bibcode:1996Natur.379...60J. doi:10.1038/379060a0
- Burton, Sir Richard Francis (1884). "The Book of the Sword". Internet archive: Chatto and Windus. p. 111. ISBN 1605204366
- Maryon, Herbert (1948). "A Sword of the Nydam Type from Ely Fields Farm, near Ely". Proceedings of the Cambridge Antiquarian Society. XLI: 73–76. doi:10.5284/1034398
- Simpson, Layne (2003). Shotguns & Shotgunning. Krause Publications. p. 256. ISBN 978-0-87349-567-7
- Ruth Bloomfeld (November 11, 2009). "Feilden Clegg Bradley's Leeds complex completed". bdonline.co.uk. Retrieved 24 September 2010
- Goddard, Wayne (2000). The Wonder of Knifemaking. Iola, WI: Krause Publications. pp. 137–145. ISBN 978-0-87341-798-3
- Maryon, Herbert (May 1960). "Pattern-Welding and Damascening of Sword-Blades—Part 2: The Damascene Process". Studies in Conservation. 5 (2): 52–60. doi:10.2307/1504953. JSTOR 1504953

2

Stainless Steel: An Overview

Stainless steel is a form of steel that does not easily rust or corrode in presence of water as normal steel does. Stainless steel can be classified into three types based on their crystalline structure. They are austenitic, martensitic and ferritic. This chapter is an overview of the subject matter incorporating all the major aspects of stainless steel.

Stainless Steel

Stainless steel cladding is used on the Walt Disney Concert Hall

In metallurgy, stainless steel, also known as inox steel or inox from French *inoxydable* (inoxidizable), is a steel alloy with a minimum of 10.5% chromium content by mass.

Stainless steel is notable for its corrosion resistance, and it is widely used for food handling and cutlery among many other applications.

Stainless steel does not readily corrode, rust or stain with water as ordinary steel does. However, it is not fully stain-proof in low-oxygen, high-salinity, or poor air-circulation environments. There are various grades and surface finishes of stainless steel to suit the environment the alloy must endure. Stainless steel is used where both the properties of steel and corrosion resistance are required.

Stainless steel differs from carbon steel by the amount of chromium present. Unprotected carbon steel rusts readily when exposed to air and moisture. This iron oxide film (the rust) is active and accelerates corrosion by making it easier for more iron oxide to form. Since iron oxide has lower density than steel, the film expands and tends to flake and fall away. In comparison, stainless steels contain sufficient chromium to undergo passivation, forming an inert film of chromium oxide on the surface. This layer prevents further corrosion by blocking oxygen diffusion to the steel surface and stops corrosion from spreading into the bulk of the metal. Passivation occurs only if the proportion of chromium is high enough and oxygen is present.

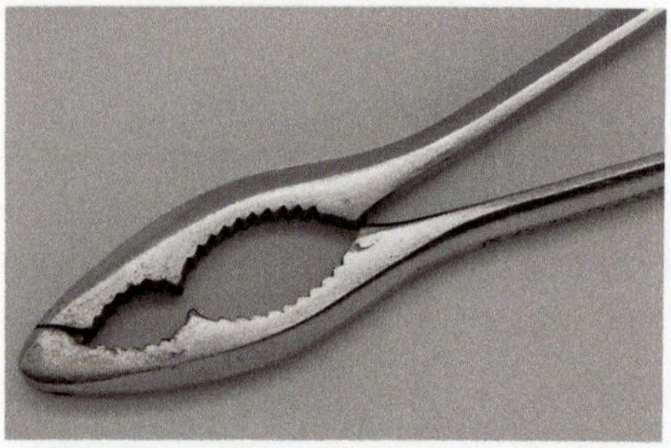

Stainless steel is used for corrosion-resistant tools such as this nutcracker

Stainless steel's resistance to corrosion and staining, low maintenance, and familiar lustre make it an ideal material for many applications. The alloy is milled into coils, sheets, plates, bars, wire, and tubing to be used in cookware, cutlery, household hardware, surgical instruments, major appliances, industrial equipment (for example, in sugar refineries) and as an automotive and aerospace structural alloy and construction material in large buildings. Storage tanks and tankers used to transport orange juice and other food are often made of stainless steel, because of its corrosion resistance. This also influences its use in commercial kitchens and food processing plants, as it can be steam-cleaned and sterilized and does not need paint or other surface finishes.

Properties

Stainless steel (row 3) resists salt-water corrosion better than aluminum-bronze (row 1) or copper-nickel alloys (row 2)

Stainless steel is not completely immune to corrosion in this desalination equipment

Oxidation

High oxidation resistance in air at ambient temperature is normally achieved with addition of a minimum of 13% (by weight) chromium, and up to 26% is used for harsh environments. The chromium forms a passivation layer of chromium(III) oxide (Cr_2O_3) when exposed to oxygen. The layer is too thin to be visible, and the metal remains lustrous and smooth. The layer is impervious to water and air, protecting the metal beneath, and this layer quickly reforms when the surface is scratched. This phenomenon is called passivation and is seen in other metals, such as aluminium and titanium. Corrosion resistance can be adversely affected if the component is used in a non-oxygenated environment, a typical example being underwater keel bolts buried in timber.

When stainless steel parts such as nuts and bolts are forced together, the oxide layer can be scraped off, allowing the parts to weld together. When forcibly disassembled, the welded material may be torn and pitted, a destructive effect known as galling. Galling can be avoided by the use of dissimilar materials for the parts forced together, for example bronze and stainless steel, or even different types of stainless steels (martensitic against austenitic). However, two different alloys electrically connected in a humid, even mildly acidic environment may act as a voltaic pile and corrode faster. Nitronic alloys, made by selective alloying with manganese and nitrogen, may have a reduced tendency to gall. Additionally, threaded joints may be lubricated to provide a film between the two parts and prevent galling. Low-temperature carburizing is another option that virtually eliminates galling and allows the use of similar materials without the risk of corrosion and the need for lubrication.

Acids

Stainless steel is generally highly resistant to attack from acids, but this quality depends on the kind and concentration of the acid, the surrounding temperature, and the type of steel. Type 904 is resistant to sulfuric acid at room temperature, even in high concentrations; types 316 and 317 are resistant below 10%; and type 304 should not be used in the presence of sulfuric acid at any concentration. All types of stainless steel resist attack from phosphoric acid, types 316 and 317 more so than 304; types 304L and 430 have been successfully used with nitric acid. Hydrochloric acid will damage any kind of stainless steel, and should be avoided.

Bases

The 300 series of stainless steel grades is unaffected by any of the weak bases such as ammonium hydroxide, even in high concentrations and at high temperatures. The same grades of stainless exposed to stronger bases such as sodium hydroxide at high concentrations and high temperatures will likely experience some etching and cracking, especially with solutions containing chlorides such as sodium hypochlorite.

Organics

Types 316 and 317 are both useful for storing and handling acetic acid, especially in solutions where it is combined with formic acid and when aeration is not present (oxygen helps protect stainless steel under such conditions), though 317 provides the greatest level of resistance to corrosion. Type 304 is also commonly used with formic acid though it will tend to discolor the solution. All grades resist damage from aldehydes and amines, though in the latter case grade 316 is preferable to 304; cellulose acetate will damage 304 unless the temperature is kept low. Fats and fatty acids only affect grade 304 at temperatures above 150 °C (302 °F), and grade 316 above 260 °C (500 °F), while 317 is unaffected at all temperatures. Type 316L is required for processing of urea.

Electricity and Magnetism

left nut is not in inox and is rusty

Poor selection of materials can cause galvanic corrosion to other metals in contact with stainless steel

Like steel, stainless steel is a relatively poor conductor of electricity, with significantly lower electrical conductivity than copper. Other metals in contact with stainless steel, particularly in a damp or acidic environment, may suffer galvanic corrosion even though the stainless metal may be unaffected.

Ferritic and martensitic stainless steels are magnetic. Annealed austenitic stainless steels are non-magnetic. Work hardening can make austenitic stainless steels slightly magnetic.

History

The corrosion resistance of iron-chromium alloys was first recognized in 1821 by French metallurgist Pierre Berthier, who noted their resistance against attack by some acids and suggested their use in cutlery. Metallurgists of the 19th century were unable to produce the combination of low carbon and high chromium found in most modern stainless steels, and the high-chromium alloys they could produce were too brittle to be practical.

In 1872, the Englishmen Clark and Woods patented an alloy that would today be considered a stainless steel.

In the late 1890s Hans Goldschmidt of Germany developed an aluminothermic (thermite) process for producing carbon-free chromium. Between 1904 and 1911 several researchers, particularly Leon Guillet of France, prepared alloys that would today be considered stainless steel.

A NON-RUSTING STEEL.

Sheffield Invention Especially Good for Table Cutlery.

According to Consul John M. Savage, who is stationed at Sheffield, England, a firm in that city has introduced a stainless steel, which is claimed to be non-rusting, unstainable, and untarishable. This steel is said to be especially adaptable for table cutlery, as the original polish is maintained after use, even when brought in contact with the most acid foods, and it requires only ordinary washing to cleanse.

"It is claimed," writes Mr. Savage in the Commerce Reports, "that this steel retains a keen edge much like that of the best double-sheer steel, and, as the properties claimed are inherent in the steel and not due to any treatment, knives can readily be sharpened on a 'steel' or by using the ordinary cleaning machine or knifeboard. It is expected it will prove a great boon, especially to large users of cutlery, such as hotels, steamships, and restaurants.

"The price of this steel is about 26 cents a pound for ordinary sizes, which is about double the price of the usual steel for the same purpose. It also costs more to work up, so that the initial cost of articles made from this new discovery, it is estimated, will be about double the present cost; but it is considered that the saving of labor to the customer will more than cover the total cost of the cutlery in the first twelve month."

An announcement, as it appeared in the 1915 *New York Times*, of the development of stainless steel in Sheffield, England.

Friedrich Krupp Germaniawerft built the 366-ton sailing yacht *Germania* featuring a chrome-nickel steel hull in Germany in 1908. In 1911, Philip Monnartz reported on the relationship between chromium content and corrosion resistance. On 17 October 1912, Krupp engineers Benno Strauss and Eduard Maurer patented austenitic stainless steel as Nirosta.

Similar developments were taking place contemporaneously in the United States, where Christian Dantsizen and Frederick Becket were industrializing ferritic stainless steel. In 1912, Elwood Haynes applied for a US patent on a martensitic stainless steel alloy, which was not granted until 1919.

Monument to Harry Brearley at the former Brown Firth Research Laboratory in Sheffield, England

In 1912, Harry Brearley of the Brown-Firth research laboratory in Sheffield, England, while seeking a corrosion-resistant alloy for gun barrels, discovered and subsequently industrialized a martensitic stainless steel alloy. The discovery was announced two years later in newspaper article in *The New York Times*. The metal was later marketed under the "Staybrite" brand by Firth Vickers in England and was used for the new entrance canopy for the Savoy Hotel in London in 1929. Brearley applied for a US patent during 1915 only to find that Haynes had already registered a patent. Brearley and Haynes pooled their funding and with a group of investors formed the American Stainless Steel Corporation, with headquarters in Pittsburgh, Pennsylvania.

In the beginning stainless steel was sold in the US under different brand names like "Allegheny metal" and "Nirosta steel". Even within the metallurgy industry the eventual name remained unsettled; in one trade journal was calling it "unstainable steel". In 1929, before the Great Depression hit, over 25,000 tons of stainless steel were manufactured and sold in the US.

Types

Pipes and fittings made of stainless steel

There are different types of stainless steels: when nickel is added, for instance, the austenite structure of iron is stabilized. This crystal structure makes such steels virtually non-magnetic and less brittle at low temperatures. For greater hardness and strength, more carbon is added. With proper heat treatment, these steels are used for such products as razor blades, cutlery, and tools.

Significant quantities of manganese have been used in many stainless steel compositions. Manganese preserves an austenitic structure in the steel, similar to nickel, but at a lower cost.

Stainless steels are also classified by their crystalline structure:

- *Austenitic*, or 200 and 300 series, stainless steels have an austenitic crystalline structure, which is a face-centered cubic crystal structure. Austenite steels make up over 70% of total stainless steel production. They contain a maximum of 0.15% carbon, a minimum of 16% chromium, and sufficient nickel and/or manganese to retain an austenitic structure at all temperatures from the cryogenic region to the melting point of the alloy.
    - 200 Series—austenitic chromium-nickel-manganese alloys. Type 201 is hardenable through cold working; Type 202 is a general purpose stainless steel. Decreasing nickel content and increasing manganese results in weak corrosion resistance.
    - 300 Series. The most widely used austenite steel is the 304, also known as *18/8* for its composition of 18% chromium and 8% nickel. 304 may be referred to as A2 stainless (not to be confused with AISI grade A2 *air hardening* alloy tool steel containing about 5% chromium). The second most common austenite steel is the 316 grade, also referred to as A4 stainless and called marine grade stainless, used primarily for its increased resistance to corrosion. A typical composition of 18% chromium and 10% nickel, commonly known as *18/10 stainless*, is often used in cutlery and high-quality cookware. *18/0* is also available.

    Superaustenitic stainless steels, such as Allegheny Ludlum alloy AL-6XN and 254SMO, exhibit great resistance to chloride pitting and crevice corrosion because of high molybdenum content (>6%) and nitrogen additions, and the higher nickel content ensures better resistance to stress-corrosion cracking versus the 300 series. The higher alloy content of superaustenitic steels makes them more expensive. Other steels can offer similar performance at lower cost and are preferred in certain applications. For example ASTM A387 is used in pressure vessels but is a low-alloy carbon steel with a chromium content of 0.5% to 9%. Low-carbon versions, for example 316L or 304L, are used to avoid corrosion problems caused by welding. Grade 316LVM is preferred where biocompatibility is required (such as body implants and piercings). The "L" means that the carbon content of the alloy is below 0.03%, which reduces the sensitization effect (precipitation of chromium carbides at grain boundaries) caused by the high temperatures involved in welding.

- *Ferritic* stainless steels generally have better engineering properties than austenitic grades, but have reduced corrosion resistance, because of the lower chromium and nickel content. They are also usually less expensive. Ferritic stainless steels have a body-centered cubic crystal system and contain between 10.5% and 27% chromium with very little nickel, if any, but some types can contain lead. Most compositions include molybdenum; some, aluminium or titanium. Common ferritic grades include 18Cr-2Mo, 26Cr-1Mo, 29Cr-4Mo, and 29Cr-4Mo-2Ni. These alloys can be degraded by the presence of chromium, an intermetallic phase which can precipitate upon welding.

Swiss Army knives are made of martensitic stainless steel.

- *Martensitic* stainless steels are not as corrosion-resistant as the other two classes but are extremely strong and tough, as well as highly machinable, and can be hardened by heat treatment. Martensitic stainless steel contains chromium (12–14%), molybdenum (0.2–1%), nickel (less than 2%), and carbon (about 0.1–1%) (giving it more hardness but making the material a bit more brittle). It is quenched and magnetic.

- *Duplex steel* stainless steels have a mixed microstructure of austenite and ferrite, the aim usually being to produce a 50/50 mix, although in commercial alloys the ratio may be 40/60. Duplex stainless steels have roughly twice the strength compared to austenitic stainless steels and also improved resistance to localized corrosion, particularly pitting, crevice corrosion and stress corrosion cracking. They are characterized by high chromium (19–32%) and molybdenum (up to 5%) and lower nickel contents than austenitic stainless steels.

  The properties of duplex stainless steels are achieved with an overall lower alloy content than similar-performing super-austenitic grades, making their use cost-effective for many applications. Duplex grades are characterized into groups based on their alloy content and corrosion resistance.

  - *Lean duplex* refers to grades such as UNS S32101 (LDX 2101), S32202 (UR2202), S32304, and S32003.

  - *Standard duplex* refers to grades with 22% chromium, such as UNS S31803/S32205, with 2205 being the most widely used.

  - *Super duplex* is by definition a duplex stainless steel with a Pitting Resistance Equivalent Number (PREN) > 40, where PREN = %Cr + 3.3x(%Mo + 0.5x%W) + 16x%N. Usually super duplex grades have 25% or more chromium. Some common examples are S32760 (Zeron 100 via Rolled Alloys), S32750 (2507), and S32550 (Ferralium 255 via Langley Alloys).

  - *Hyper duplex* refers to duplex grades with a PRE > 48. UNS S32707 and S33207 are the only grades currently available on the market.

- *Precipitation-hardening martensitic* stainless steels have corrosion resistance comparable to austenitic varieties, but can be precipitation hardened to even higher strengths than the other martensitic grades. The most common, 17-4PH, uses about 17% chromium and 4% nickel.

The designation "CRES" is used in various industries to refer to corrosion-resistant steel. Most mentions of CRES refer to stainless steel, although the correspondence is not absolute, because there are other materials that are corrosion-resistant but not stainless steel.

Grades

There are over 150 grades of stainless steel, of which 15 are most commonly used. There are a number of systems for grading stainless and other steels, including US SAE steel grades.

Comparison of Standardized Steels

EN-standard Steel no. k.h.s DIN	EN-standard Steel name	SAE grade	UNS
1.4512	X6CrTi12	409	S40900
		410	S41000
1.4016	X6Cr17	430	S43000
1.4109	X65CrMo14	440A	S44002
1.4112	X90CrMoV18	440B	S44003
1.4125	X105CrMo17	440C	S44004
		440F	S44020
1.4310	X10CrNi18-8	301	S30100
1.4318	X2CrNiN18-7	301LN	
1.4301	X5CrNi18-10	304	S30400
1.4307	X2CrNi18-9	304L	S30403
1.4306	X2CrNi19-11	304L	S30403
1.4311	X2CrNiN18-10	304LN	S30453
1.4948	X6CrNi18-11	304H	S30409
1.4303	X5CrNi18-12	305	S30500
	X5CrNi30-9	312	
1.4841	X22CrNi2520	310	S31000
1.4845	X 5 CrNi 2520	310S	S31008
1.4401	X5CrNiMo17-12-2	316	S31600
1.4408	G-X 6 CrNiMo 18-10	316	S31600
1.4436	X3CrNiMo17-13-3	316	S31600

EN-standard Steel no. k.h.s DIN	EN-standard Steel name	SAE grade	UNS
1.4406	X2CrNiMoN17-12-2	316LN	S31653
1.4404	X2CrNiMo17-12-2	316L	S31603
1.4432	X2CrNiMo17-12-3	316L	S31603
1.4435	X2CrNiMo18-14-3	316L	S31603
1.4571	X6CrNiMoTi17-12-2	316Ti	S31635
1.4429	X2CrNiMoN17-13-3	316LN	S31653
1.4438	X2CrNiMo18-15-4	317L	S31703
1.4541	X6CrNiTi18-10	321	S32100
1.4878	X12CrNiTi18-9	321H	S32109
1.4362	X2CrNi23-4	2304	S32304
1.4462	X2CrNiMoN22-5-3	2205	S31803/S32205
1.4501	X2CrNiMoCuWN25-7-4	J405	S32760
1.4539	X1NiCrMoCu25-20-5	904L	N08904
1.4529	X1NiCrMoCuN25-20-7		N08926
1.4547	X1CrNiMoCuN20-18-7	254SMO	S31254

Standard Finishes

316L stainless steel, with an unpolished, mill finish

Standard mill finishes can be applied to flat rolled stainless steel directly by the rollers and by mechanical abrasives. Steel is first rolled to size and thickness and then annealed to change the properties of the final material. Any oxidation that forms on the surface (mill scale) is removed by pickling, and a passivation layer is created on the surface. A final finish can then be applied to achieve the desired aesthetic appearance.

- No. 0: Hot rolled, annealed, thicker plates
- No. 1: Hot rolled, annealed and passivated
- No. 2D: Cold rolled, annealed, pickled and passivated
- No. 2B: Same as above with additional pass through highly polished rollers
- No. 2BA: Bright annealed (BA or 2R) same as above then bright annealed under oxygen-free atmospheric condition
- No. 3: Coarse abrasive finish applied mechanically
- No. 4: Brushed finish
- No. 5: Satin finish
- No. 6: Matte finish (brushed but smoother than #4)
- No. 7: Reflective finish
- No. 8: Mirror finish
- No. 9: Bead blast finish
- No. 10: Heat colored finish—offering a wide range of electropolished and heat colored surfaces

Applications

The 630-foot-high (190 m), stainless-clad (type 304) Gateway Arch defines St. Louis's skyline

The pinnacle of New York's Chrysler Building is clad with *Nirosta* stainless steel, a form of Type 302

An art deco sculpture on the Niagara-Mohawk Power building in Syracuse, New York

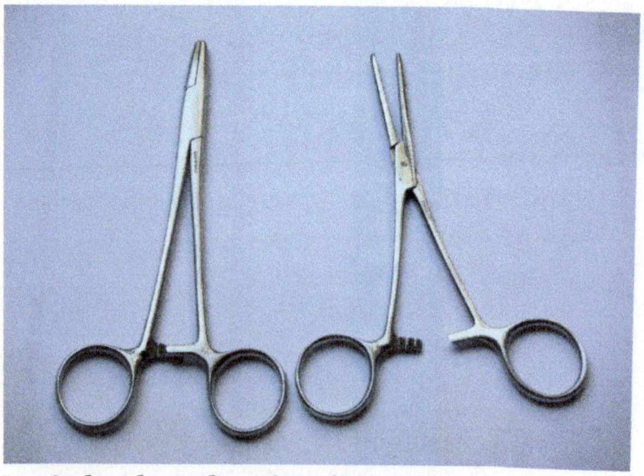

Surgical tools, such as these hemostats, are commonly made of stainless steel

Stainless steel is often used for everyday kitchen sinks, appliances, and cookware

Stainless steel is often used for cookware

Architecture

Stainless steel is used for buildings for both practical and aesthetic reasons. Stainless steel was in vogue during the art deco period. The most famous example of this is the upper portion of the Chrysler Building (pictured). Some diners and fast-food restaurants use large ornamental panels and stainless fixtures and furniture. Because of the durability of the material, many of these buildings still retain their original appearance. Stainless steel is used today in building construction because of its durability and because it is a weldable building metal that can be made into aesthetically pleasing shapes. An example of a building in which these properties are exploited is the Art Gallery of Alberta in Edmonton, which is wrapped in stainless steel.

Type 316 stainless is used on the exterior of both the Petronas Twin Towers and the Jin Mao Building, two of the world's tallest skyscrapers.

The Parliament House of Australia in Canberra has a stainless steel flagpole weighing over 220 tonnes (240 short tons).

The aeration building in the Edmonton Composting Facility, the size of 14 hockey rinks, is the largest stainless steel building in North America.

Bridges

- The Helix Bridge is a pedestrian bridge linking Marina Centre with Marina South in the Marina Bay area in Singapore.
- Cala Galdana Bridge in Minorca (Spain) was the first stainless steel road bridge.
- Sant Fruitos Pedestrian Bridge (Catalonia, Spain), arch pedestrian bridge.
- Padre Arrupe Bridge (Bilbao, Spain) links the Guggenheim museum to the University of Deusto.

Monuments and Sculptures

- Unisphere, constructed as the theme symbol of the 1964 New York World's Fair, is constructed of Type 304L stainless steel as a spherical framework with a diameter of 120 feet (37 m) (New York City)
- Gateway Arch (pictured) is clad entirely in stainless steel: 886 tons (804 metric tonnes) of 0.25 in (6.4 mm) plate, #3 finish, type 304 stainless steel. (St. Louis, Missouri)
- United States Air Force Memorial has an austenitic stainless steel structural skin (Arlington, Virginia)
- Atomium was renovated with stainless-steel cladding in a renovation completed in 2006; previously the spheres and tubes of the structure were clad in aluminium (Brussels, Belgium)
- Cloud Gate sculpture by Anish Kapoor (Chicago, Illinois)
- Sibelius Monument is made entirely of stainless steel tubes (Helsinki, Finland)
- The Kelpies (Falkirk, Scotland)
- Man of Steel (sculpture) under construction (Rotherham, England)
- Juraj Jánošík monument (Terchova, Slovakia)

Airports

Stainless steel is a modern trend for roofing material for airports due to its low glare reflectance to keep pilots from being blinded, also for its properties that allow thermal reflectance in order to keep the surface of the roof close to ambient temperature. The Hamad International Airport in Qatar was built with all stainless steel roofing for these reasons, as well as the Sacramento International Airport in California.

Locomotion

Automotive Bodies

The Allegheny Ludlum Corporation worked with Ford on various concept cars with stainless steel bodies from the 1930s through the 1970s to demonstrate the material's potential. The 1957 and 1958 Cadillac Eldorado Brougham had a stainless steel roof. In 1981 and 1982, the DeLorean DMC-12 production automobile used Type-304 stainless steel body panels over a glass-reinforced plastic monocoque. Intercity buses made by Motor Coach Industries are partially made of stainless steel. The aft body panel of the Porsche Cayman model (2-door coupe hatchback) is made of stainless steel. It was discovered during early body prototyping that conventional steel could not be formed without cracking (due to the many curves and angles in that automobile). Thus, Porsche was forced to use stainless steel on the Cayman.

Some automotive manufacturers use stainless steel as decorative highlights in their vehicles.

Passenger Rail Cars

Rail cars have commonly been manufactured using corrugated stainless steel panels (for additional structural strength). This was particularly popular during the 1960s and 1970s, but has since declined. One notable example was the early Pioneer Zephyr. Notable former manufacturers of stainless steel rolling stock included the Budd Company (USA), which has been licensed to Japan's Tokyu Car Corporation, and the Portuguese company Sorefame. Many railcars in the United States are still manufactured with stainless steel, unlike other countries who have shifted away.

Aircraft

Budd also built two airplanes, the Budd BB-1 Pioneer and the Budd RB-1 Conestoga, of stainless steel tube and sheet. The first, which had fabric wing coverings, is on display at the Franklin Institute, being the longest continuous display of an aircraft ever, since 1934. The RB-2 Was almost all stainless steel, save for the control surfaces. One survives at the Pima Air & Space Museum, adjacent to Davis–Monthan Air Force Base.

The American Fleetwings Sea Bird amphibious aircraft of 1936 was also built using a spot-welded stainless steel hull.

Due to its thermal stability, the Bristol Aeroplane Company built the all-stainless steel Bristol 188 high-speed research aircraft, which first flew in 1963. However, the practical problems encountered meant that Concorde employed aluminium alloys.

The use of stainless steel in mainstream aircraft is hindered by its excessive weight compared to other materials, such as aluminium.

Medicine

Surgical tools and medical equipment are usually made of stainless steel, because of its durability and ability to be sterilized in an autoclave. In addition, surgical implants such as bone reinforcements and replacements (e.g. hip sockets and cranial plates) are made with special alloys formulated to resist corrosion, mechanical wear, and biological reactions *in vivo*.

Stainless steel is used in a variety of applications in dentistry. It is common to use stainless steel in many instruments that need to be sterilized, such as needles, endodontic files in root canal therapy, metal posts in root canal–treated teeth, temporary crowns and crowns for deciduous teeth, and arch wires and brackets in orthodontics. The surgical stainless steel alloys (e.g., 316 low-carbon steel) have also been used in some of the early dental implants.

Culinary Use

Stainless steel is often preferred for kitchen sinks because of its ruggedness, durability, heat resistance, and ease of cleaning. In better models, acoustic noise is controlled by applying resilient undercoating to dampen vibrations. The material is also used for cladding of surfaces such as appliances and backsplashes.

Cookware and bakeware may be clad in stainless steels, to enhance their cleanability and durability, and to permit their use in induction cooking. Because stainless steel is a poor conductor of heat, it is often used as a thin surface cladding over a core of copper or aluminum, which conduct heat more readily.

Cutlery is normally stainless steel, both for low corrosion, ease of cleaning, negligible toxicity, as well as not flavoring the food by electrolytic activity.

Jewelry

Stainless steel is used for jewelry and watches, with 316L being the type commonly used for such applications. It can be re-finished by any jeweler and will not oxidize or turn black.

Valadium, a stainless steel and 12% nickel alloy is used to make class and military rings. Valadium is usually silver-toned, but can be electro-plated to give it a gold tone. The gold tone variety is known as Sun-lite Valadium. Other "Valadium" types of alloy are trade-named differently, with such names as "Siladium" and "White Lazon".

Firearms

Some firearms incorporate stainless steel components as an alternative to blued or parkerized steel. Some handgun models, such as the Smith & Wesson Model 60 and the Colt M1911 pistol, can be made entirely from stainless steel. This gives a high-luster finish similar in appearance to nickel plating. Unlike plating, the finish is not subject to flaking, peeling, wear-off from rubbing (as when repeatedly removed from a holster), or rust when scratched.

3D Printing

Some 3D printing providers have developed proprietary stainless steel sintering blends for use in rapid prototyping. One of the more popular stainless steel grades used in 3D printing is 316L stainless steel. Due to the high temperature gradient and fast rate of solidification, stainless steel products manufactured via 3D printing tend to have a more refined microstructure; this in turn results in better mechanical properties. However, stainless steel is not used as much as materials like Ti6Al4V in the 3D printing industry; this is because manufacturing stainless steel products via traditional methods is currently much more economically competitive.

Maintenance

If treated or stored incorrectly, any grade of stainless steel may discolor or stain. To maintain optimum appearance, the surface should be cared for regularly.

The quality of installation affects the durability and lifespan of stainless steel. Therefore, it is important to make sure stainless steel is in good condition before installation. Normally, giving it a quick clean is enough prior to installation. However, if surface contamination is present, more attention is required. In fields such as aerospace, pharmaceuticals and food handling, an extremely high standard of cleanliness may be required, so extra care should be taken.

Maintenance is required to maintain the quality and appearance of steel. Depending on the environment, it is carried out between one and ten times per year. A proper maintenance routine significantly prolongs the life of stainless steel.

Abrasive cleaning tools should be avoided to prevent alteration of stainless steel finishes. Chloride-containing solutions, such as bleach, should also be avoided.

- Soft cloth and water: suitable for cosmetic issues and general cleaning
- Mild detergent: needed if stains cannot be easily lifted with water
- Glass cleaner: useful for removing fingerprints and similar stains

Despite its design and use, stainless steel can still be susceptible to corrosion, some grades more than others, and especially in corrosive environments. Challenging environments include saline environments, such as coastal areas where regular exposure to sea salt is common and areas where de-icing salts are common during winter. Manufacturing environments, especially in chemical and food industries, may also be subject to corrosive substances.

Stainless steel may also corrode if surfaces come into direct contact with iron or carbon steel. Trace particles from iron or carbon steel will rust on stainless steel surfaces. If left unattended, rust spots may compromise surface passivation and may spread internally. Contamination is common when stainless steel is subject to sparks from nearby welding, cutting, drilling, or grinding of carbon steel.

Corrosion can be treated by:

- Light rust: all-purpose lubricant or domestic stainless steel cleaners (typically containing calcium carbonate or citric acid)
- Moderate rust: phosphorus acid solutions
- Severe rust: hydrofluoric acid bath (typically performed by professional service providers due to the hazardous nature of chemicals)

Recycling and Reusing

Stainless steel is 100% recyclable. An average stainless steel object is composed of about 60% recycled material of which approximately 40% originates from end-of-life products and about

60% comes from manufacturing processes. According to the International Resource Panel's Metal Stocks in Society report, the per capita stock of stainless steel in use in society is 80–180 kg in more developed countries and 15 kg in less-developed countries.

There is a secondary market that recycles usable scrap for many stainless steel markets. The product is mostly coil, sheet, and blanks. This material is purchased at a less-than-prime price and sold to commercial quality stampers and sheet metal houses. The material may have scratches, pits, and dents but is made to the current specifications.

Health Effects

Stainless steel is generally considered to be biologically inert, but some sensitive individuals develop a skin irritation due to a nickel allergy caused by certain alloys.

Stainless steels contain typically 10-30 % chromium besides other elements like C, Mn, Si, S etc. Chromium imparts corrosion resistance to steel. Varying amounts of other alloying elements like Ni, Mo,

V, Ti, Ni, etc may be added to obtain certain specific property. There are different types of stainless steels like

Austenitic stainless steels: which contain 18% Cr, 8% Ni, and C is in between 0.03-0.15%

Ferritic stainless steels: which contain 12% to 30% Cr and 0.08% to 0.12%C.

Martensitic stainless steels: which contain around 13% Cr and C varying in between 0.15% to 0.25%. Certain grades contain C 0.6% to 0.95%.

Duplex stainless steels: in which Cr is around 25%.

Precipitation hardenable stainless steel: contain 18-20% Cr, 8 to 10 % Ni and copper, Titanium, Aluminum.

It may be noted that all stainless steels contain chromium and carbon besides other elements. Production of stainless steels requires controlling chromium and carbon.

Thermodynamics of Decarburization of Chromium Melt

In stainless steel making both chromium and carbon oxidizes when decarburization of melt is done. The Ellingham diagram for oxide formation indicates that carbon oxidation in preference to chromium oxidation can occur at temperatures greater than 1220°C, when both elements are in pure state. Under all practical conditions carbon oxidation can occur at temperatures above 1800°C in presence of chromium.

Chromium oxidizes to $p_{co} = \text{atlm}$ and C oxidizes to CO.

$$2C + O_2 = 2CO$$

$$2Cr + 1.5\,O_2 = Cr_2O_3$$

$$3Cr + 2O_2 = Cr_3O_4$$

Equilibrium distribution between Cr and C is represented by considering Cr_3O_4

$$Cr_3O_4 + 4C = 3Cr + 4\{CO\}$$

Assuming pure Cr_3O_4 and Cr and using henry's law for carbon, it follows

$$K = \frac{[f_{cr}w_{cr}]^3 P_{co}^4}{[f_c w_c]^4}$$

$$\frac{[w_{cr}]^3}{[w_c]^4} = \frac{K \times (f_c)^4}{f_{cr}^3 p_{co}^4}$$

f_c and f_{cr} are activity coefficient of carbon and chromium in liquid iron at 1wt % standard state. Hilly and Kaveney proposed the following equation for distribution of chromium and carbon. The equation above suggests that decrease in p_{co} can increase the ratio $\frac{[w_{cr}]^3}{[w_c]^4}$

$$\log\left[\frac{w_{cr}}{w_c}\right] = -\frac{13800}{T} + 8.76 - 0.925\log p_{co}$$

The effect of Ni

$$\log\left[\frac{w_{cr}}{w_c}\right] = -\frac{13800}{T + 4.21wt\%Ni} + 8.76 - 0.925\log p_{co}$$

Equations above describes distribution ratio of Cr and c in Fe-Ni –Cr-C melt. Table shows the influence of temperature and pCO on the distribution ratio $(R) = \left[\frac{w_{cr}}{w_c}\right]$ without Ni and $R^1 = \left[\frac{w_{cr}}{w_c}\right]$ in the presence of 10Wt % Ni.

Table: chromium /carbon distribution ratio

		$p_{co} = atlm$		$p_{co} = 0.25atm$		$p_{co} = 0.1atm$
T(k)	R	R¹	R	R¹	R	R¹
1873	25	36	89	129	207	301
1973	55	82	209	295	488	690
2073	128	173	460	619	1077	1477
2173	240	339	863	1220	2019	2852

We can infer the following from the table:

1. Increase in temperature increases R and R^1 at all p_{co} values. High temperature is required to suppress chromium oxidation in favor of carbon oxidation.

2. Decrease in p_{co} increases both R and R^1 at all temperatures.

3. Nickel increases R and R^1

The above observations suggest that high temperature is needed to remove carbon in presence of chromium, if stainless steel is produced at atmospheric pressure. If reduced pressures are used; lower temperature can cause oxidation of carbon.

Figure shows carbon and chromium contents of simple Fe-Cr-C melt in equilibrium with pure solid Cr_2O_3 and CO gas at one atmospheric pressure at various temperatures. Carbon in the melt can be preferentially oxidized at much higher temperature without much loss of chromium. This can be done by oxygen lancing to molten bath. Any excess chromium would be oxidized and will have to be recovered from slag.

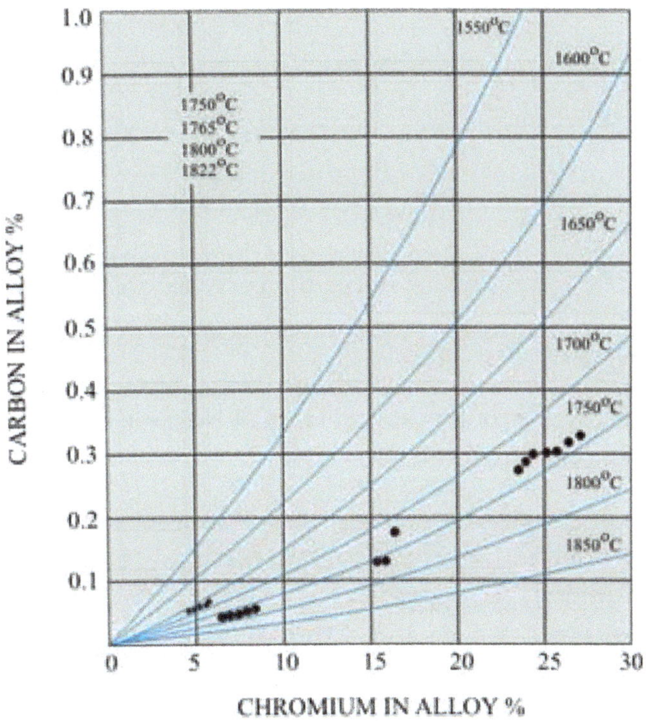

Variation of carbon content with chromium at different temperatures under equilibrium conditions

For a fix value of temperature T and partial pressure of CO, decreasing carbon content in the melt decreases the equilibrium chromium level. Hence, decarburization of the melt is not possible without chromium oxidation.

Therefore, thermodynamic alternative is to lower the equilibrium partial pressure of CO gas of the melt, so that further decarburization take place in absence of chromium oxidation at a specific temperature as shown in figure. With the increasing temperature and decreasing partial pressure of CO gad, equilibrium condition can be shifted to lower carbon levels. The injection of inert gas (argon or nitrogen) lowers the partial pressure of CO gas in the bath, thus allowing higher chromium contents to be in equilibrium with much lower carbon content.

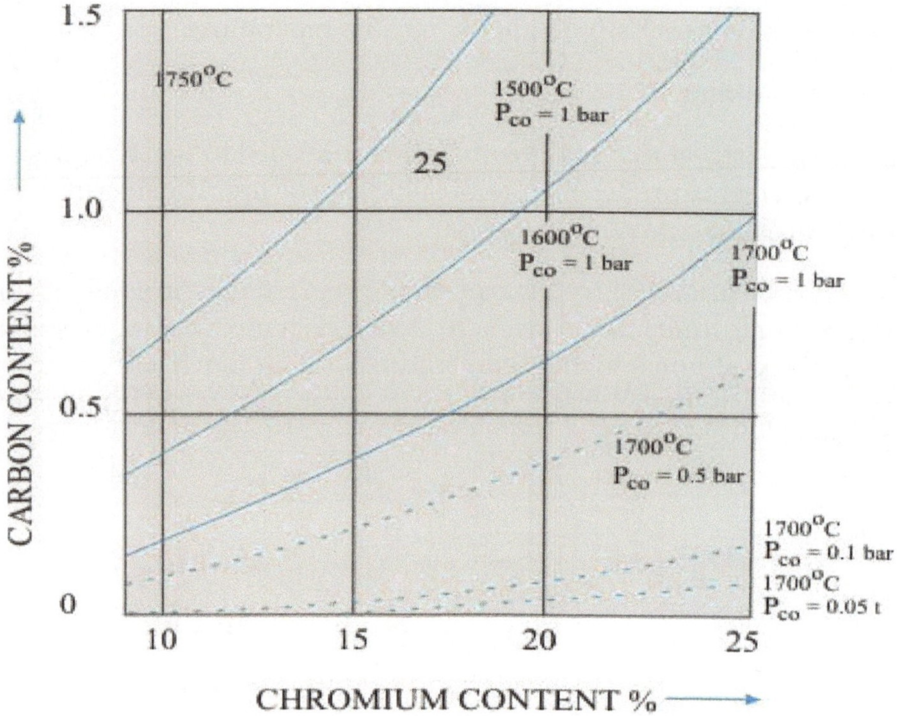

Variation of carbon content with chromium at different temperatures and partial presssures of CO

Technology of Stainless Steel Making

Electric arc furnace was used to produce stainless steel by melting scrap of the desired composition. EAF was used only as a melting unit.

Typically, charge consists of carbon steel scrap + stainless steel scrap +lime. The charge is melted in EAF and after melt- down period, the melt contains around 10% Cr, all Nickel and carbon. Melt consists of Fe- Cr –Ni –C alloy.

Oxygen is blown onto Fe- Cr- Ni –C melt and basic Cr_2O_3 slag forms. Initially chromium oxidizes until bath temperature rises to 1800°C . Carbon oxidation occurs once the bath temperature rises to 1800°C In the finishing stage, low carbon ferrochrome is added to make the chromium content of steel to a desired value. The disadvantages with this technology are:

- High temperature is required which cause damage to the refectory lining.
- Low carbon ferrochrome is required which in expensive.

Basis of Development of New Technology

Thus a technology which can use high carbon ferro- chrome and to decarburize the bath at selectively lower temperatures would be required. If carbon is to be oxidized at low temperatures in preference to Cr , a reduction in pressure of CO from 1 atmosphere to lower value would be required. Reduction in pressure of CO can be achieved either by vacuum or by using a mixture of $Ar + O_2$. The former one is vacuum oxygen decarburization (VOD) and the later is (AOD).

AOD Process

The process in carried out in a converter type of vessel. Vessel is lined with magnesite brick. A mixture of argon + oxygen is injected through the tuyeres located on the side of the converter shell.

Fe-Cr- Ni-C alloy melt is prepared in EAF. Melt is charged in AOD vessel. High carbon-ferrochrome is charged. In the initial stage a mixture of O_2 : Ar in 3:1 ratio is blown through the side tuyeres. When carbon reduces to 30% of the original value the ratio of O_2 : Ar is changed to 2: 1 and blow is continued to attain 0.09 to 0.12% C. First stage of blow generates sufficient amount of heat due to oxidation of Cr and hence coolants are added (5% of the change). Stainless steel scrap is used.

In the final stage, the ratio of O_2 : Ar is changed 1:3 to bring C to the desired value. Fe- Si is added to recover Cr from slag and slag basicity is maintained at 1.5 to 2 by adding lime.

$$Cr_3O_4 + 2Si = 3Cr + 2SiO_2,$$

Slag formation and slag metal reactions are facilitated by argon stirring of the bath. The bath is desulphurized to levels well below 0.015%.

Austenitic Stainless Steel

Austenitic stainless steel is a specific type of stainless steel alloy. Stainless steels may be classified by their crystalline structure into three main types: *austenitic*, *ferritic* and *martensitic*.

300 Series austenitic stainless steel has austenite as its primary phase (face centered cubic crystal). These are alloys containing chromium and nickel, and sometimes molybdenum and nitrogen, structured around the Type 302 composition of iron, 18% chromium, and 8% nickel. 200 Series austenitic stainless steels replace the majority of their nickel content with manganese to reduce cost. Austenitic steels are not hardenable by heat treatment. The most familiar stainless steel is probably SAE 304 stainless steel, also called 18/8 or A2 stainless steel. Type 304 surgical stainless steel is an austenitic steel containing 18–20% chromium and 8–10% nickel.

Alloy 20 (Carpenter 20) is an austenitic stainless steel possessing excellent resistance to hot sulfuric acid and many other aggressive environments which would readily attack type 316 stainless. This alloy exhibits superior resistance to stress-corrosion cracking in boiling 20–40% sulfuric acid. Alloy 20 has excellent mechanical properties and the presence of niobium in the alloy minimizes the precipitation of carbides during welding.

Austenitic stainless steel can be tested by nondestructive testing using the dye penetrant inspection method but not the magnetic particle inspection method. Eddy-current testing may also be used. It is the primary stainless steel used in aviation construction. Chemical and steel industry-applicable grades are SAE steel grades 308, 308L, 316, 316L, 316LN (nitrogen bearing), 312, 309L, and 310L. "L" denotes a carbon percentage less than 0.03%, mostly used for corrosion heat resistance and reducing sensitivity to chromium carbide formation. Another grade, 312, is used for dissimilar steel welding, also known as universal alloy steel as unknown composition steels can be welded. For high temperature applications at greater than 600°C, grades 309 and 310 are preferred. Grade 316 is alloyed with molybdenum (~2–3%) for high-temperature strength, pitting and crevice corrosion resistance.

Martensitic Stainless Steel

Martensitic stainless steel is a specific type of stainless steel alloy.

Stainless steels may be classified by their crystalline structure into three main types: *austenitic*, *ferritic* and *martensitic*. Martensitic stainless steels can be high- or low-carbon steels built around the Type 410 composition of iron, 12% chromium, and up to 1.2% carbon. They are usually tempered and hardened. Tempered martensite gives steel good hardness and high toughness; used largely for medical tools (scalpels, razors and internal clamps). Untempered martensite is low in toughness and therefore brittle.

The characteristic body-centered tetragonal martensite microstructure was first observed by German microscopist Adolf Martens around 1890. In 1912, Elwood Haynes applied for a U.S. patent on a martensitic stainless steel alloy. This patent was not granted until 1919.

Martensitic stainless steel can be nondestructively tested using the magnetic particle inspection method, unlike austenitic stainless steel.

Also in 1912, Harry Brearley of the Brown-Firth research laboratory in Sheffield, England, while seeking a corrosion-resistant alloy for gun barrels, discovered and subsequently industrialized a martensitic stainless steel alloy. The discovery was announced two years later in newspaper article in *The New York Times*. Brearly applied for a U.S. patent during 1915. This was later marketed under the "*Staybrite*" brand by Firth Vickers in England and was used for the new entrance canopy for the Savoy Hotel in 1929 in London.

References

- Akhavan Tabatabae, Behnam; et al. (2009). "Influence of Retained Austenite on the Mechanical Properties of Low Carbon Martensitic Stainless Steel Castings". ISIJ International. 51 (3): 471–475. doi:10.2355/isijinternational.51.471
- Ashby, Michael F.; David R. H. Jones (1992) [1986]. "Chapter 12". Engineering Materials 2 (with corrections ed.). Oxford: Pergamon Press. p. 119. ISBN 0-08-032532-7
- "The Recycling of Stainless Steel ("Recycled Content" and "Input Composition" slides)" (Flash). International Stainless Steel Forum. 2006. Retrieved 19 November 2006
- Carlisle, Rodney P. (2004) Scientific American Inventions and Discoveries, p. 380, John Wiley and Sons, ISBN 0-471-24410-4
- Wu, Wenjie; Maye, Mathew M. (2014-01-01). "Void Coalescence in Core/Alloy Nanoparticles with Stainless Interfaces". Small. 10 (2): 271–276. ISSN 1613-6829. doi:10.1002/smll.201301420
- Anusavice, Kenneth J. (2003) Phillips' Science of Dental Materials, 11th Edition. W.B. Saunders Company. ISBN 0721693873. p. 639

Modern Steelmaking Practices

3

There has been a growing need to have ecologically-viable technology to produce steel in terms of available resources, energy consumption and impact on environment. Two methods that follow these principles are energy optimizing furnace and CONARC. This chapter discusses in detail the theories and methodologies related to modern steelmaking practices.

Energy Optimizing Furnace

- To develop ecological balance technology. Such a technology would require processing the raw materials such that liquid products are discharged at the environmental temperature and waste products are discharged in their harmless states. This requires integrating the concept of recovery, reuse and recirculation within the process environment.
- To develop a process which is flexible to use varying proportion of metallic materials depending on their availability. Converter steelmaking is based on hot metal.
- To develop technologies which are not electric energy dependent. Electric steelmaking is dependent of availability of cheap electric power, though developments are made to use chemical energy.

In the above concept, two technologies are worthwhile to mention: Energy optimizing furnace (EOF) and CONARC (Converter + Arc furnace).

Energy Optimizing Furnace (EOF)

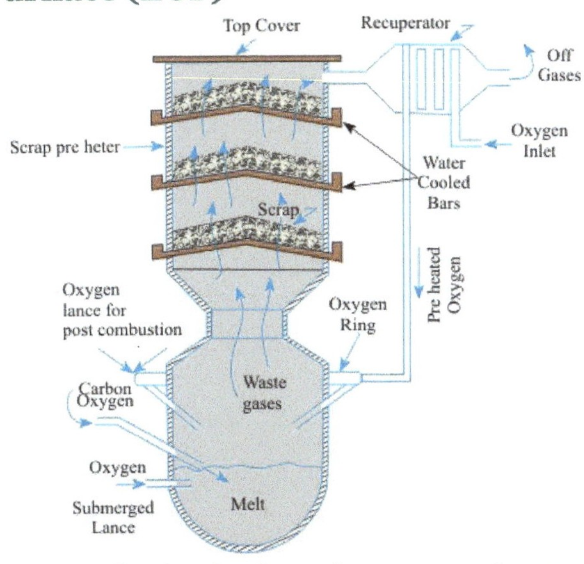

Arrangement of furnace, scrap preheating chamber and a recuperator in an Energy optimizing Furnace

Figure is a schematic sketch of an energy optimizing furnace (EOF). EOF is a combination of three independent, interconnected reactors, namely furnace to produce steel, preheater to heat the scrap and a recuperator to rcover waste heat and to reuse by heating oxygen.

Design of each reactor and their integration with each other is important such that furnace off gas exits the system at a low temperature to the extent possible and at the same time scrap is heated to the desired temperature and steel of required quality is tapped.

Furnace

The furnace is refractory lined and has a provision for injection of carbon and oxygen. Oxygen can be injected both at high speed to promote decarburization and at low speed to promote post combustion.

Preheater

A preheater is installed at the top of furnace to preheat the scrap by the waste gases flowing upward from the furnace. For scrap preheating residence time (R T) of the waste gas in the preheater

$$RT = \frac{\text{Diameter of the pre heater}}{\text{flow velocity of exit gas}} = 0.785\frac{D^3}{Q}$$

should be maximum. D is diameter of preheater and Q is volume flow rate of furnace gases. Both D and Q are important. For a given Q, diameter of preheater must be such that furnace gases can cool within $700\,°C$ to $800\,°C$ from $1300\,°C$ to $1350\,°C$. The exit temperature of gas from the preheater is the inlet temperature of gas to the recuperator. Scrap size is also important to heat scrap to the desired temperature. Smaller size scrap can be preheated efficiently.

Recuperator

Recuperator is a counter current heat exchanger in which hot gases at $700\,°C$ to $800\,°C$ from the scrap preheater enters and exit gases leave at $350\,°C$ and oxygen enters at $25\,°C$ and preheated oxygen enters the furnace. The efficiency of the heat exchanger depends on the flow velocity of hot gas and oxygen. For a given cross section of the flow passage of the recuperator, the length of the recuperator must be optimized so that heat transfer from the furnace gases is maximum to preheat the incoming oxygen. The efficiency of the recuperator can be defined as

$$\in = \frac{\text{sensible heat in oxygen at the preheat temperature}}{\text{sensible heat in furnace gases}}$$

Furnace Operation

A liquid heel is left in the furnace from the previous heat in proportion which depends on scrap proportion. Higher the scrap proportion, higher is the amount of heel.

The operation in this case starts with injection of carbon in the hot metal heel until 3% carbon is achieved. This is followed by charging of the preheated scrap and refining commences immediately thereafter. Oxygen injection for refining and post combustion begins simultaneously. Slag is allowed to form so that it can flow over the sill down the pit.

The refractory consumption is kept within the limits by using water cooled panels and scrap supporting bass. To increase furnace life and effective preheating, maximum and minimum size of scrap are controlled.

Important Features

- Oxygen is blown by a lance submerged in the melt for refining and simultaneously oxygen is also injected for post combustion.
- Utilization of sensible heat of furnace off-gases by preheating scrap and O_2
- Flexibility to use hot metal and scrap in any proportion
- Wet de-dusting system
- Liquid steel with low levels of phosphorus and sulphur
- Low noise level
- An average of 42 heat/ day is possible.

With the above features EOF presents a low cost alternative for steelmaking combined with flexibility in terms of metallic charge mix.

Steel Quality Aspects

1. Due to continuous slag flushing, turndown P can be obtained up to 0.008% and sulphur up to 0.025%.
2. High % hot metal produces a product with very low tramp elements, Advantages for die forging steels, special clean steels and steels for seamless piper etc.
3. High CO partial pressure during the blow leads to very low N and H content in steel at turndown.

Basic Characteristics

Standard capacities of EOF are available in capacities 30/40 t, 60/ 80 t and 100/ 120 t.

Hearth surface is 6.6 to 22 square meters

Shell diameter is around 5.3 to 7.5m

Total height from working platform to top level =17 to 25 m

No. of scrap-preheater staged 1 to 2

Tilting tapping angle 8^0

Main Equipments Description

1. Bottom car (2unit) is used for quick bottom exchange for a new campaign. One bottom car supports the EOF for operation and the other supports a second bottom for relining. Tilting is allowed at high speed hydraulic cylinders to allow slag free tapping.

2. Gas cleaning plant is wet system.

3. EOF has facilities for charging and weighing of bulk material and ferro-alloys.

Advantages

1. Flexibility to in terms of metallic charge
2. No need of electric energy
3. High productivity
4. Low tramp elements
5. Low inclusions due to slag free toppings

Energy Saving is Due to

1. Post combustion (95%)
2. Scrap preheating ($850^0 C$)
3. High operational of flexibility.

Consumption of Raw Materials/ ton of Liquid Steel

Hot metal and scrap+ pig iron are charged in 70% :30% proportion

Lime 45 kg/t (depending on P content of HM)

O_2 is consumed at 50-70 N cu.m./t

Fuel consumption is 5-10 M cal/t

Tapping- ($1700^0 C$) without ladle furnace and ($1650^0 C$) with ladle furnace.

CONARC

In seventies there were discussions amongst scientists and technologies about the future of pure top blown and pure bottom blown steelmaking. These discussions resulted into a combined top and bottom blown steelmaking technology which has almost replaced pure top blown and pure bottom blown steel making.

In recent years discussions are in progress about EAF and oxygen steelmaking with respect to use of energy and raw material.

The objective is to develop a technology utilizing benefits of both EAF and top blown steelmaking. CONARC is the resultant of such discussions.

What is CONARC?

The CONARC steelmaking process has been developed by Mannesman Demag. It combines electrode arc melting with the oxygen steelmaking process. The process consists of two shells with one 3 phase electrode arm and one top blowing lance for hot metal treatment.

This allows a high flexibility with respect to use of raw materials as well as energy source. World-wide there are around 12 units in operation.

- In India Ispat industries in Dolvi, Maharashtra has been integrated with Ispat's converter and arc furnace. This combination has given flexibility of alternative choices of energy resources to use for steelmaking-be it coal, electricity or gas. In addition, choice of alterative metallic charge material is also available i.e. be it iron are, pellets, sponge iron, scrap or hot metal

This plant will be equipped with compact strip production technology to produce high grade finished steel like hot rolled coils.

Bhushan steel and strips limited have also installed CONARC.

Recuperator

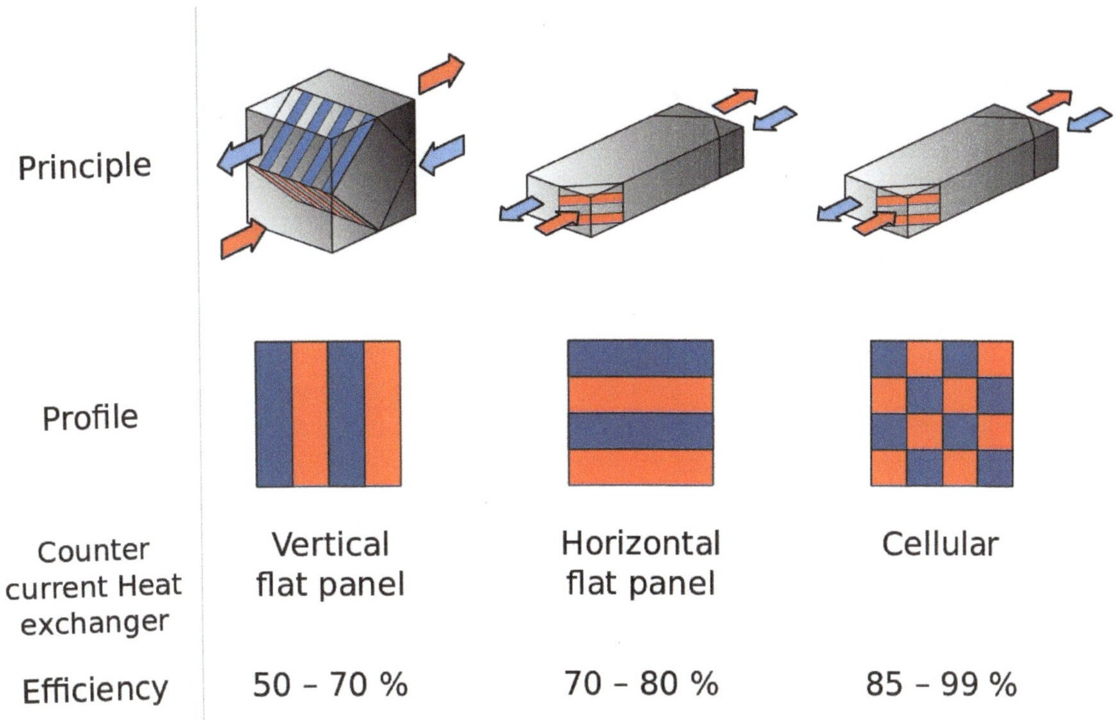

Types of recuperator, or cross plate heat exchanger

A recuperator is a special purpose counter-flow energy recovery heat exchanger positioned within the supply and exhaust air streams of an air handling system, or in the exhaust gases of an industrial process, in order to recover the waste heat.

Description

In many types of processes, combustion is used to generate heat, and the recuperator serves to recuperate, or reclaim this heat, in order to reuse or recycle it. The term recuperator refers as well to liquid-liquid counterflow heat exchangers used for heat recovery in the chemical and refinery industries and in closed processes such as ammonia-water or LiBr-water absorption refrigeration cycle.

Recuperators are often used in association with the burner portion of a heat engine, to increase the overall efficiency. For example, in a gas turbine engine, air is compressed, mixed with fuel, which is then burned and used to drive a turbine. The recuperator transfers some of the waste heat in the exhaust to the compressed air, thus preheating it before entering the fuel burner stage. Since the gases have been pre-heated, less fuel is needed to heat the gases up to the turbine inlet temperature. By recovering some of the energy usually lost as waste heat, the recuperator can make a heat engine or gas turbine significantly more efficient.

Energy Transfer Process

Normally the heat transfer between airstreams provided by the device is termed as "sensible heat", which is the exchange of energy, or enthalpy, resulting in a change in temperature of the medium (air in this case), but with no change in moisture content. However, if moisture or relative humidity levels in the return air stream are high enough to allow condensation to take place in the device, then this will cause "latent heat" to be released and the heat transfer material will be covered with a film of water. Despite a corresponding absorption of latent heat, as some of the water film is evaporated in the opposite airstream, the water will reduce the thermal resistance of the boundary layer of the heat exchanger material and thus improve the heat transfer coefficient of the device, and hence increase efficiency. The energy exchange of such devices now comprises both sensible and latent heat transfer; in addition to a change in temperature, there is also a change in moisture content of the exhaust air stream.

However, the film of condensation will also slightly increase pressure drop through the device, and depending upon the spacing of the matrix material, this can increase resistance by up to 30%. If the unit is not laid to falls, and the condensate not allowed to drain properly, this will increase fan energy consumption and reduce the seasonal efficiency of the device.

Use in Ventilation Systems

In heating, ventilation and air-conditioning systems, HVAC, recuperators are commonly used to re-use waste heat from exhaust air normally expelled to atmosphere. Devices typically comprises a series of parallel plates of aluminium, plastic, stainless steel, or synthetic fiber, alternate pairs of which are enclosed on two sides to form twin sets of ducts at right angles to each other, and which contain the supply and extract air streams. In this manner heat from the exhaust air stream is transferred through the separating plates, and into the supply air stream.

Manufacturers claim gross efficiencies of up to 80% depending upon the specification of the unit.

The characteristics of this device are attributable to the relationship between the physical size of the unit, in particular the air path distance, and the spacing of the plates. For an equal air pressure drop through the device, a small unit will have a narrower plate spacing and a lower air velocity than a larger unit, but both units may be just as efficient. Because of the cross-flow design of the unit, its physical size will dictate the air path length, and as this increases, heat transfer will increase but pressure drop will also increase, and so plate spacing is increased to reduce pressure drop, but this in turn will reduce heat transfer.

As a general rule a recuperator selected for a pressure drop of between 150–250 pascals (0.022–0.036 psi) will have a good efficiency, while having a small effect on fan power consumption, but will have in turn a higher seasonal efficiency than that for physically smaller, but higher pressure drop recuperator.

When heat recovery is not required, it is typical for the device to be bypassed by use of dampers arranged within the ventilation distribution system. Assuming the fans are fitted with inverter speed controls, set to maintain a constant pressure in the ventilation system, then the reduced pressure drop leads to a slowing of the fan motor and thus reducing power consumption, and in turn improves the seasonal efficiency of the system.

Use in Metallurgical Furnaces

Recuperators have also been used to recover heat from waste gasses to preheat combustion air and fuel for many years by metallic recuperators to reduce energy costs and carbon footprint of operation. Compared to alternatives such as regenerative furnaces, initial costs are lesser, there are no valves to be switching back and forth, there are no induced-draft fans and it does not require a web of gas ducts spread up all over the furnace.

Historically the recovery ratios of recuperators compared to regenerative burners were low. However, recent improvements to technology have allowed recuperators to recover 70-80% of the waste heat and pre-heated air up to 850–900 °C (1,560–1,650 °F) is now possible.

Use in Microturbines

Recuperators can be used to increase the efficiency of gas turbines for power generation, provided the exhaust gas is hotter than the compressor outlet temperature. The exhaust heat from the turbine is used to pre-heat the air from the compressor before further heating in the combustor, reducing the fuel input required. The larger the temperature difference between turbine out and compressor out, the greater the benefit from the recuperator. Therefore, microturbines (<1MW), which typically have low pressure ratios, have the most to gain from the use of a recuperator. In practice, a doubling of efficiency is possible through the use of a recuperator. The major practical challenge for a recuperator in microturbine applications is coping with the exhaust gas temperature, which can exceed 750 °C (1,380 °F).

Finery Forge

A finery forge is a hearth used to fine (i.e., produce, refine) wrought iron, through the decarburization of the pig iron. The fining process involved liquifying cast iron in a fining hearth and removing carbon from the molten cast iron through oxidation. Finery forges were used as early as 3rd century BC, based on archaeological evidence found at a site in Tieshengguo, China. The finery forge process was replaced by puddling process and the roller mill, both developed by Henry Cort in 1783-4, but not becoming widespread until after 1800.

Hearth (left) and trip hammer (centre) in a finery forge.
In the back room (right) is a large pile of charcoal.

History

A finery forge was used to refine wrought iron at least by the 3rd century BC in ancient China, based on the earliest archaeological specimens of cast and pig iron fined into wrought iron and steel found at the early Han Dynasty (202 BC – 220 AD) site at Tieshengguo. Pigott speculates that the finery forge existed in the previous Warring States period (403–221 BC), because of the wrought iron items from China dating to that period and there was no documented evidence of the bloomery ever being used in China. Wagner writes that in addition to the Han Dynasty hearths believed to be fining hearths, there is also pictoral evidence of the fining hearth from a Shandong tomb mural dated 1st to 2nd century AD, as well as a hint of written evidence in the 4th century AD Daoist text *Taiping Jing*.

In Europe, the concept of the finery forge may have been evident as early as the 13th century. However, it was perhaps not capable of being used to fashion plate armor until the 15th century, as described in conjunction with the waterwheel-powered blast furnace by the Florentine Italian engineer Antonio Averlino (c. 1400 - 1469). The finery forge process began to be replaced in Europe from the late 18th century by others, of which puddling was the most successful, though some continued in use through the mid-19th century. The new methods used mineral fuel (coal or coke), and freed the iron industry from its dependence on wood to make charcoal.

Types

Interior of the preserved Walloon forge in Österbybruk, Uppland

German Forge

The dominant type in Sweden was the German forge, which had a single hearth that was used for all processes.

Walloon Forge

In Swedish Uppland north of Stockholm and certain adjacent provinces, another kind known as the Walloon forge was used, mainly for the production of a particularly pure kind of iron known as oregrounds iron, which was exported to England to make blister steel. Its purity depended on the use of ore from the Dannemora mine. The Walloon forge was virtually the only kind used in Great Britain.

The forge had two kinds of hearths, the finery to finish the product and the chafery to reheat the bloom that was the raw material of the process.

Lancashire Forge

Process

In the finery, a workman known as the "finer" remelted pig iron so as to oxidise the carbon (and silicon). This produced a lump of iron (with some slag) known as a bloom. This was consolidated using a water-powered hammer and returned to the finery.

The next stages were undertaken by the "hammerman", who in some iron-making areas such as South Yorkshire was also known as the "stringsmith", who heated his iron in a string-furnace. Because the bloom is highly porous, and its open spaces are full of slag, the hammerman's or stringsmith's tasks were to beat (work) the heated bloom with a hammer to drive the molten slag out of it, and then to draw the product out into a bar to produce what was known as anconies or bar iron. In order to do this, he had to reheat the iron, for which he used the chafery. The fuel used in the finery had to be charcoal (later coke), as impurities in any mineral fuel would affect the quality of the iron.

Slag

The waste product was allowed to cool in the hearth and removed as a "mosser". In the Furness district they were often left as the capstone of a wall, particularly near Spark Bridge and Nibthwaite forges.

Cementation Process

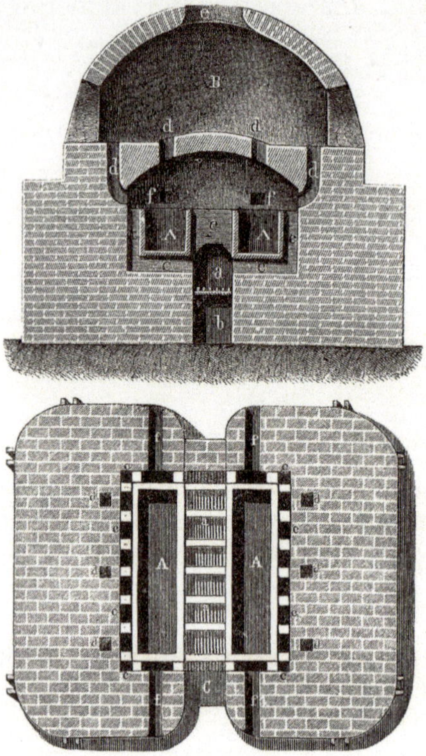

Fig. 302 et 303. — Four de cémentation anglais.

Cementation furnace

The cementation process is an obsolete technology for making steel by carburization of iron. Unlike modern steelmaking, it increased the amount of carbon in the iron. It was apparently developed before the 17th century. Derwentcote Steel Furnace, built in 1720, is the earliest surviving example of a cementation furnace. Another example in the UK is the cementation furnace in Doncaster Street, Sheffield.

The process was described in a treatise published in Prague in 1574. It was again invented by Johann Nussbaum of Magdeburg, who began operations at Nuremberg (with partners) in 1601. The process was patented in England by William Ellyot and Mathias Meysey in 1614. At that date, the "invention" could consist merely of the introduction of a new industry or product, or even a mere monopoly. They evidently soon transferred the patent to Sir Basil Brooke, but he was forced to surrender it in 1619. A clause in the patent prohibiting the import of steel was found to be undesirable because he could not supply as much good steel as was needed. Brooke's furnaces were probably in his manor of Madeley at Coalbrookdale (which certainly existed before the English Civil War) where two cementation furnaces have been excavated. He probably used bar iron from the Forest of Dean, where he was a partner in farming the King's ironworks in two periods.

By 1631, it was recognised that Swedish iron was the best raw material and then or later particularly certain marks (brands) such as *double bullet* (so called from the mark OO) from Österby and *hoop L* from Leufsta (now Lövsta), whose mark consisted of an L in a circle, both belonging to Louis De Geer and his descendants. These were among the first ironworks in Sweden to use the Walloon process of fining iron, producing what was known in England as oregrounds iron. It was so called from the Swedish port of Öregrund, north of Stockholm, in whose hinterland most of the ironworks lay. The ore used came ultimately from the Dannemora mine.

Process

The Doncaster Street cementation furnace in Sheffield, England

The process begins with wrought iron and charcoal. It uses one or more long stone *pots* inside a furnace. Typically, in Sheffield, each was 14 feet by 4 feet and 3.5 feet deep. Iron bars and charcoal are packed in alternating layers, with a top layer of charcoal and then refractory matter to make the pot or "coffin" airtight. Some manufacturers used a mix of powdered charcoal, soot and mineral salts, called *cement powder*. In larger works, up to 16 tons of iron was treated in each cycle,though it can be done on a small scale, such as in a small furnace or blacksmith's forge.

Depending on the thickness of the iron bars, the pots were then heated from below for a week or more. Bars were regularly examined and when the correct condition was reached the heat was withdrawn and the pots were left until cool—usually around fourteen days. The iron had "gained" a little over 1% in mass from the carbon in the charcoal, and had become heterogeneous bars of *blister steel.*

The bars were then shortened, bound, heated and forge welded together to become *shear steel.*It would be cut and re welded multiple times, with each new weld producing a more homogeneous, higher quality steel. This would be done at most 3-4 times, as more is unnecessary and could potentially cause carbon loss from the steel.

Alternatively they could be broken up and melted in a crucible using a crucible furnace with a flux to become *crucible steel* or *cast steel*, a process devised by Benjamin Huntsman in Sheffield in the 1740s.

Similar Processes

In the early modern period, brass, an alloy of copper and zinc, was usually produced by a cementation process in which metallic copper was heated with calamine, a zinc ore, to make calamine brass.

Converter Steelmaking

Converter steelmaking processes are also known as BOF (Basic oxygen Furnace) steelmaking. These processes include top blown steelmaking and combined blown steelmaking processes. These processes are based on hot metal. Hot metal from Blast furnace is refined to steel. Blast furnace hot metal contains 3-4 % C, 0.8 to 1% Si, 0.6 to 0.8 % Mn and 0.15 - 0.2 % P. In all BOF processes, oxygen is blown from top. In most of the steelmaking practices hot metal is pretreated to remove Si, P and S from hot metal to the extent it is possible.

Pretreatment of Hot Metal

In recent years pretreatment of hot metal prior to charging in converter has become common practice. The objective is to reduce S, P and Si of hot metal to produce steel with good surface finish, free from internal cracks. Here S removal is discussed.

Removal of Sulphur

Removal of sulphur is called desulphurization. Sulphur exhibits negative deviation from Henry's Law in molten iron. Bonds between Fe and S are strong. Activity coefficient of S increases with increase in C, Si and P.

Desulphurization reaction is a slag /metal reaction.

$$[S]+(O^{2-})=(S^{2-})+[O]$$

$$\frac{a_{S^{2-}}}{[Wt\%S]}=\frac{(Wt\%S)}{[Wt\%S]}=\frac{K\,f_s a_{O^{2-}}}{[Wt\%O]}$$

K is equilibrium constant of reaction above and f_s is activity coefficient of sulphur in metal.

$(Wt\%S)/[Wt\%S]$ is called partition coefficient of sulphur or index of desulphurization. From equation 2 the best conditions for desulphurization can be derived.

- High f_s i.e. high activity coefficient of sulphur in metal. Carbon, silicon and phosphorus increases f_s which means hot metal is better for desulphurization than molten steel.

- High $a_{O^{2-}}$ in slag which means highly basic slag.
- Low $[Wt\%O]$ in metal. Reaction above suggests that removal of S from hot metal is accompanied by oxygen transfer to metal. This oxygen must be removed for efficient desulphurization.
- High temperature which leads to high K.

Desiliconization and Dephosphorization

Dephosphorization requires low temperature, basic slag and oxidizing conditions. Presence of silicon in hot metal impedes dephosphorization.

Reagents for Hot Metal Pretreatment

- Soda ash

It is effective reagent for both desulphurization and desiliconization. But disposal of soda bearing slag is a problem. Also soda ash generates dense fumes on addition to hot metal.

- Calcium carbide and Magnesium granules

Both are highly efficient desulphurizing agents and can decrease sulphur content to very low value. Calcium carbide along with lime is injected into the bath. Magnesium granules help to reduce injection time and slag volume.

- In some situations mill scale and sinter fines are used to desiliconize hot metal.

Location of Hot Metal Pretreatment

Hot metal pretreatment can be carried out either at blast furnace runner, or in transfer ladle or in torpedo. Pretreatment in the blast furnace runner has certain advantages: such as adequate mixing of the reagents due to flowing of hot metal. This practice saves time and increases ladle availability compared with when treatment is carried out in ladles.

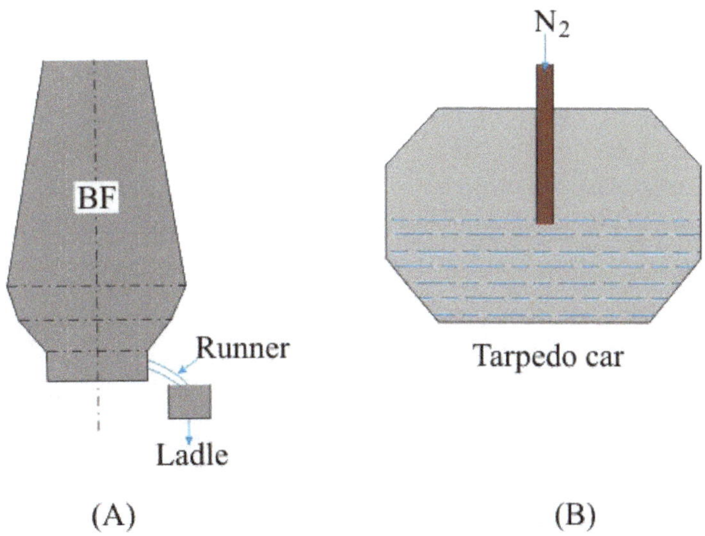

(a) Blast furnace runner (b) Torpedo car

Material Balance

Hot metal of composition 0.8% Si, 0.2% P, 0.25% Mn, 4% C and rest iron is refined in a converter to produce steel of composition 0.1% C and rest iron. Pure oxygen is blown. During refining scrap is charged whose amount is 15% of hot metal. The composition of slag is CaO 54%, FeO18%, and MnO2.5% with CaO $|SiO_2 = 3.5$. Exit gases analyses $15\%CO_2$ and $85\%CO_2$. Calculate amount of steel, slag, oxygen and waste gases per ton hot metal.

Basis: 1000 kg hot metal.

Fe balance: Let X kg is mass of steel and Y kg mass of slag

Mn Balance:

Mn in hot metal = Mn in slag

Mn in hot metal = Mn in slag
Putting the values in 3 and 4 we get.
X = 1080.68 kg steel
Y = 126.7 kg slag.

Carbon Balance:

C in hot metal + C in scrap = C in steel + C in waste gases

C in waste gases = 39.07 kg

C in $CO_2 = 5.86 kg$

C in $CO = 33.21 kg$

Exit gas volume $= 73.93 m^3$ (1 atmospheric pressure and 273K).

O_2 required can be calculated from Si, C, Mn and Fe balance.

Calculations give $O_2 = 52.70 m^3$ (1atmospheric pressure and 273K)

Material balance gives an idea of the charge materials required.

For example 100 ton capacity would require approximately 93 tons hot metal and 10 ton scrap. The amount of slag would be 12 tons and total oxygen required would be $1900 m^3$. The volume of exit gas is $6800 m^3$. These are the approximate values and are given to develop a feel of the converter design.

Design of Converter

From the metallurgical point of view an ideal converter keeps the liquid steel in space and allows all necessary metallurgical reactions to take place within the temperature range of $1400-1600^o C$.

The mechanical part, which keeps the liquid steel in space, is steel shell lined with refractory material.

The inner volume enclosed by the refractory should be maximized so as to achieve an optimum metallurgical process without sloping of slag. A ratio of 3 m^3 (internal volume/ ton) is typical in converter design.

A modern LD-converter consists of a top cone with the lip ring a barrel section and a lower cone with a dished bottom. The nose diameter and angle are chosen with reference to problems of heat loss, erosion, skulling, and stability of nose lining.

The vessel is supported by a suspension system which transmits the load to the trunion ring.

Converter design requires knowing height of molten steel bath, (h_b) diameter of bath (d_b) working height of the converter, (h_w) as shown in the figure.

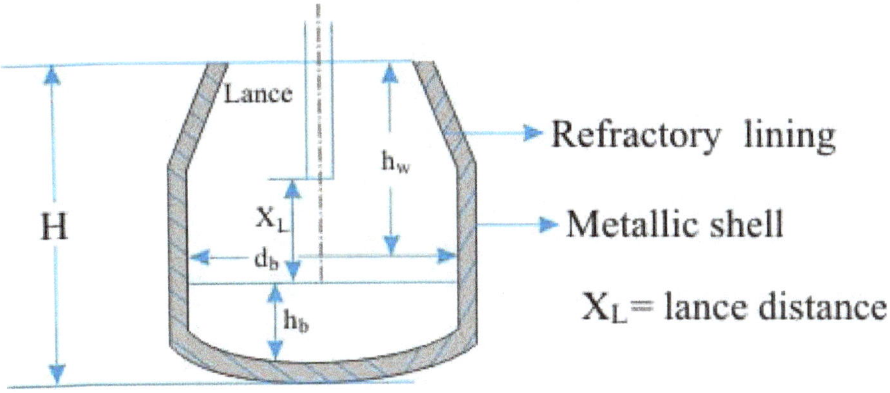

Nomenclature of the bath dimensions of a converter

Some correlations are given below:

$$\frac{h_b}{d_b} = 0.328(T)^{-0.0148}$$

T is capacity in tons, and

$$d_b\,(m) = 0.704(T)^{-0.386}$$

$$\frac{\dot{V}}{\text{min.tons}} = 3.07$$

Where $\dot{V}$ is m^3 of oxygen

For 150 ton converter capacity

$$h_b = 1.48\text{m}, d_b = 4.87\text{m}, \dot{V} = 460\frac{\text{m}^3}{\text{min}}(1\text{atm}, 273\text{K}).$$

$$h_W \approx 3.5 \times \text{bath height} = 5.2\,\text{m}$$

Total height of converter is 6.7 m excluding bottom refractory thickness.

Assuming bottom refractory thickness to be around 1 to 1.5 m, total converter height from conical to bottom becomes approximately 8m.

Lance

Oxygen gas is supplied through a water cooled lance, the tip of which fitted with multi-hole Laval nozzles made of copper.

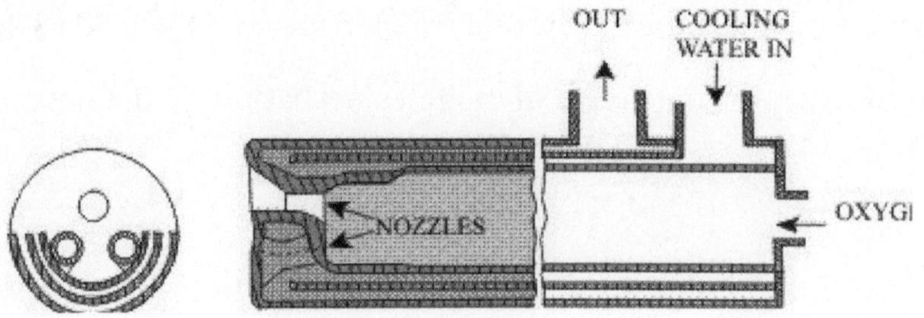

Lance is nearly 8-10m long and its diameter varies between 20cm to 25cm depending on the furnace capacity. Water requirements are around $50-70\ m^3/hr$ at a pressure of $5-7\ kg/cm^3$. Lance movement is controlled by electrically operated gear system.

Lance life is determined by the life of the nozzles. Failures of the lance may be due to faulty cooling, manufacturing defects and differential expansion between copper tip and steel tube.

The most important part of the lance is the Laval nozzle. Functions of the nozzle are:

- Supply and distribution of oxygen
- To produce a gaseous jet
- To induce bath agitation,
- To produce droplets

Lance is designed to produce non-coalescing free oxygen get. The lance is designed to operate at an upstream pressure of 10- 12 bar.

Shop Layout

Layout requires rational arrangement of equipments to ensure smooth handling of solid raw materials like scrap, fluxes, movement of oxygen lance and hot metal. Also layout should ensure smooth flow of ladles containing hot metal and steel. The refining of hot metal to steel is very fast and hence an efficient system of material transport and weighing is required.

Some essential considerations of layout are:

- A tall shop is required to raise and lower the lance in the vessel
- An elaborate gas cleaning facilities are required.
- The number of vessels in a shop may be generally two or three and one out of two or two out of three operating at a time.

- The refractory lining maintenance facilities must be adequate.
- An efficient process control strategy using computers and automatic spectro-chemical analytical methods is required.
- The shop is provided with separate cranes for handling hot metal, refined steel, scrap and slag.
- The hot metal mixer should be located on the ground floor.

Feed Materials:

- Hot metal
- Cold pig iron
- Steel scrap
- Fluxes
- Gaseous oxygen

Hot Metal

Sulphur in the hot metal should be close to final specification level.

Silicon content of hot metal determines amount of lime and slag.

Mn in hot metal produces MnO. MnO tends to retard the dephosphorization of the bath. Mn could be 0.5 to 0.8%.

Temperature of hot metal at charging is around $1250°C$ to $1300°C$. Proportion of hot metal in the charge is 75-90%.

Fluxes

Commonly used lime/limestone/dolomite to bring down the softening point of the oxides, to reduce the viscosity of slag and to decrease the activity of some components to make them stable in the slag phase.

Scrap and Ore

Used as coolants to best utilize the excess heat energy. Iron ore is sometimes used by some plant as a coolant and to promote slag formation.

Oxygen

Consumption of oxygen per tonne of steel varies with proportion of scrap and ore, and also with single and multi-hole designs and is approximately $2.5-3\ m^3/min$.

Source of heat: No external heat. Refining reactions are exothermic.

Deoxidisers and Alloying Elements

Elements like Al, ferrosilicon and ferromanganese are added as deoxidisers. The elements like Cr, Ni V, Nb etc are added as alloying elements. Carbon is added to recarburize steel if required.

Refining of Hot Metal

After the previous heat is tapped and slag is drained, lining is inspected. Scrap and hot metal are charged. Converter is tilted into the vertical position and the lance is lowered in the vessel to start the blowing. Selection of the starting lance distance is such that the concentration of the force at the bath level should not cause ejection of tiny iron particles (sparking) and at the same time maximum bath surface area is covered by the oxygen jet. The starting lance distance (X_i) for specific oxygen blowing rate $3\frac{Nm^3}{ton \times min}$ can be calculated by

$$X_i = 0.541(d_b)^{1.04}$$

d_b is bath diameter in meter. For 150 Tons converter, $d_b = 4.87\,m$ and $X_i = 2.8m$, when oxygen flow rate is approximately $450\ Nm^3/min$..

Initially oxygen is blown soft by keeping lance distance higher to promote slag formation and to avoid ejection of small particles, because hot metal is not covered by slag. Lime may be added either at the beginning of the blow or in portion during the blow. Oxygen is blown for nearly 15-20 minutes by progressively decreasing the lance distance such that slag foaming remains under control and oxidation reactions occur uninterruptedly. Slag and metal samples are analyzed.

Composition and Temperature During the Blow

Typical variation of composition of metal with the blowing time is shown in figure.

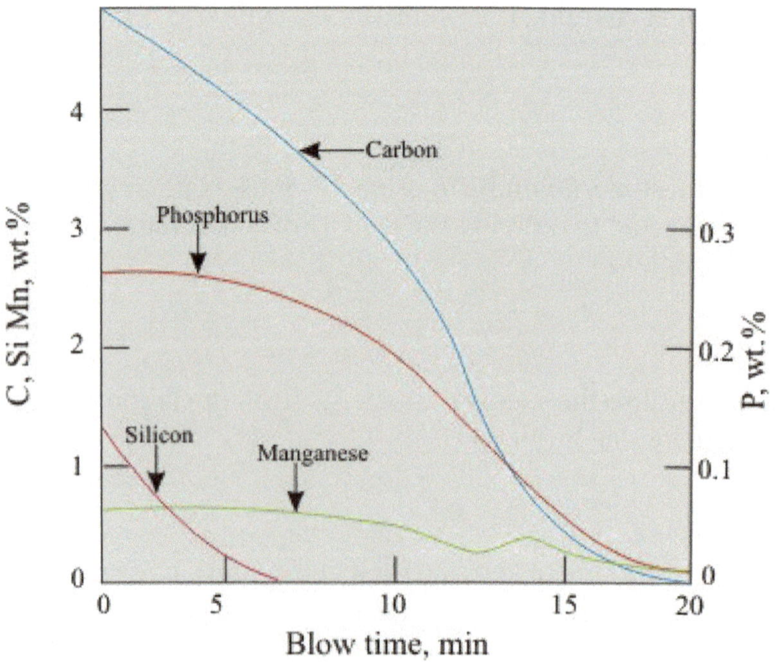

Variation of composition of metal during the blow

The following observations can be made:

- Impurities like C, Si, Mn begin to oxidize simultaneously. Si and Mn oxidize faster relative to C. Also Fe oxidizes to FeO. Rate of carbon removal is low in the beginning.
- Formation of slag begins with the oxidation of $Si - SiO_2, Mn \rightarrow MnO$ and $Fe \rightarrow Feo$
- Dissolution of lime increases during the blow. In the initial periods FeO helps lime dissolution.
- Formation of basic and limy slag promotes removal of P. It may be noted that once slag is formed both C and P removal occur simultaneously.
- In the initial stages, carbon removal rate is kept lower than P removal since P removal is favoured at lower temperatures. If carbon is removed at a faster rate in the beginning bath temperature would increase which impedes dephosphorization. Once phosphorus removal is complete, carbon removal rate can be increased.
- Note that Mn content of metal decreases initially but at later periods of blow Mn content of bath increases. This is due to the onset of the following reaction:

$$(MnO)+[Fe]=[Mn]+(FeO)$$

In the later stages of the blow bath temperature increases due to decrease in carbon content and at the sametime FeO content of slag decreases. Both conditions are responsible for increase in Mn content of the bath. To overcome, sometimes iron ore additions are made to increase the FeO content of slag to adjust the Mn content of steel.

- Temperature of the bath increases continuously.

Physico-chemical Interactions

Physico- chemical interactions of molten bath with oxygen jet depends on the lance profile i.e. change of lance height during the blow. The lance profile is specific to each converter and depends on converter profile, hot metal composition, oxygen flow rate, hot metal chemistry and steel of desired composition. Nevertheless, in all converters initial lance distance is such as to promote iron oxidation so that dissolution of CaO commences. The idea is to create a basic and limy slag at the early part of the blow to onset dephosphorization. Shallow jet penetration covers the larger bath surface and is favorable more for iron and silicon oxidation. Small amount of carbon may be removed. Once slag is formed, lance is lowered. Oxygen jet penetrates into the bath and carbon reaction favours because oxygen is available now deep into the bath. At the same time, force of the oxygen jet creates metal droplets and as a consequence three phase dispersion of gas/slag /metal droplets are formed which enhance the rate of decarburization. Figure shows the lance profile and the accompanying physico-chemical interactions.

We note that at higher lance distance, oxygen jet penetration into bath is shallow and slag formation occurs. As the lance distance is decreased, jet penetrates deep into the bath, carbon reaction commences, CO forms, droplets are produced which together leads to the formation of a three phase dispersion consisting of gas bubbles/slag/metal droplets. In this state of blow, both carbon and phosphorus removal occur at a faster rate. Formation of three phase dispersion is a characteristic feature of the top blown steelmaking. Three phase dispersion creates conditions for faster removal rates of C and P. Foaming of slag has to be controlled to avoid expulsion of slag, which can be controlled by controlling C reaction with FeO in slag. Reaction between C and FeO of slag in slag will not allow CO bubbles to grow. Smaller size gas bubbles can be trapped easily in slag as compared to larger sizes. Slag may foam and may be expelled from the converter.

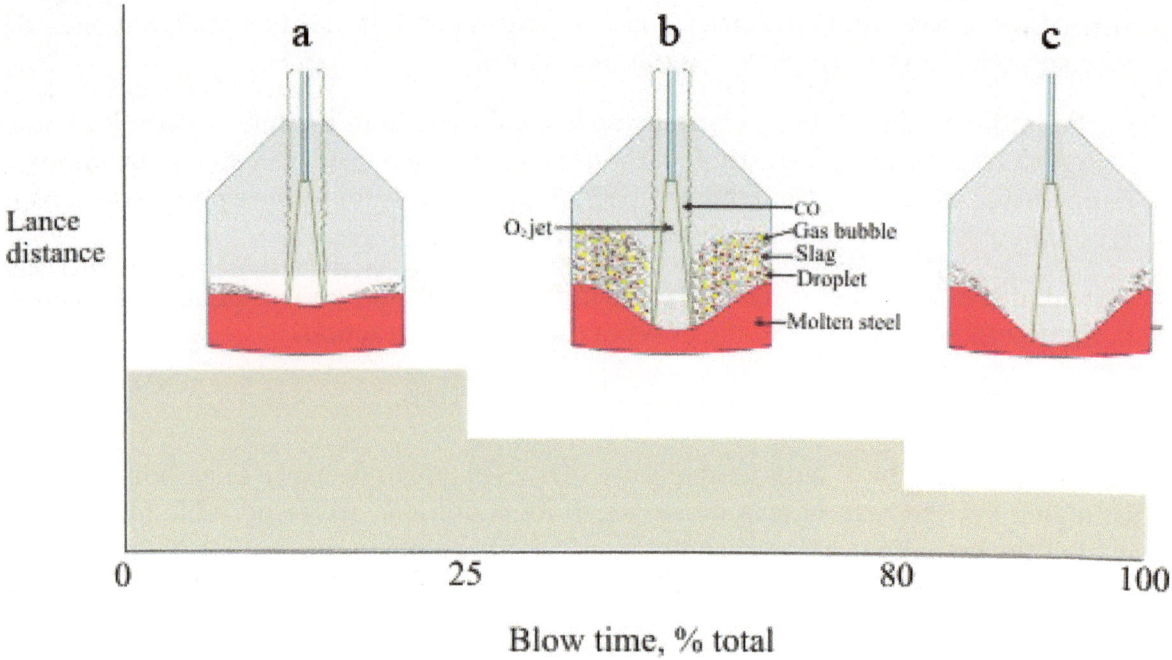

Variation of lance distance with blow time alongwith physico-chemical reactions in the converter.

Developments in Top Blown Steelmaking Practice

The most important development in top blown steelmaking practice is the simultaneous gas stirring of the bath form the bottom of the converter. This has resulted in combination blowing processes. These processes differ in terms of bottom gas rate, number and arrangement of bottom tuyeres and type of bottom injection elements i.e. porous plugs or tuyeres and whether inert gas or oxidizing gas is used.

All processes which use top blowing of oxygen and bottom stirring by inert gas is known as bath stirred top blown processes. In another type, oxygen is blown form top and bottom and is called top and bottom blowing processes.

Concept of Bottom Stirring in Top Blowing

In pure top blown steelmaking, bath agitation is very weak particularly during the initial and final stages of the blow. In the initial stages Si and Mn removal delays carbon removal whereas in the final stages carbon removal rate decreases. Figure shows decarburization rate vs time, typically observed in top blown steelmaking practice.

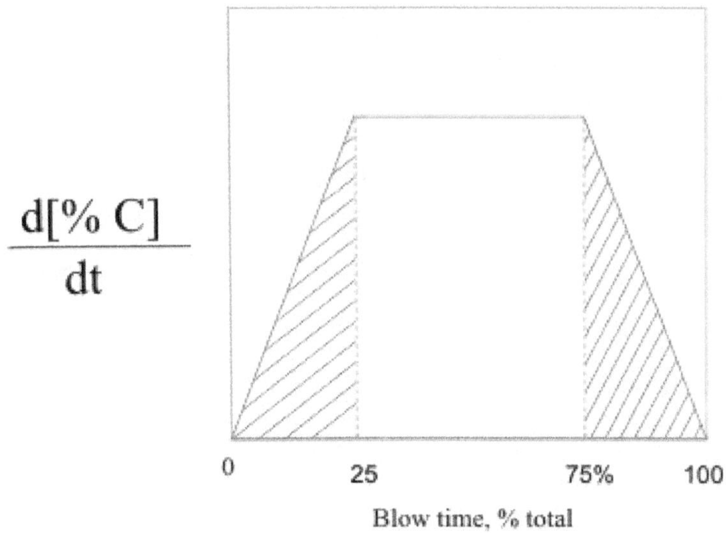

Rate of carbon removal as a function of time of blow

In the hatched regions, CO evolution in the bath is very low in pure top blown steelmaking; O_2 jet could not produce adequate bath stirring. Evolution of CO is the principle cause of bath agitation. Both in initial periods (silicon oxidation period) and in final periods (where rate of carbon removal is mass transfer controlled), evolution of CO is low. Slag analysis reveals higher rate of oxidation of Fe to FeO in both the periods which is due to weak stirring in the bath. In main part of the blow higher carbon removal rate produces higher amount of CO and produces enough bath stirring. It is considered appropriate to introduce bottom stirring gas in a top blowing converter to stir the bath.

Top blowing attributes

- Energetic supply of oxygen
- Control of slag formation
- Control of oxygen distribution.
- Simultaneous removal of C and P.
- Inadequate stirring of slag /metal phases.

In bottom blown steelmaking, all oxygen is injected through the bottom tuyeres .Though this technology provides efficient bath stirring and enhanced carbon removal, but it is difficult to distribute oxygen within the bath and also to control slag formation.

As compared with top blowing, bottom blowing changes somewhat the course of reaction of carbon, manganese, phosphorous oxidation, resulting in a different degree of oxidation of bath.

Bottom blowing increases sharply the intensity of bath stirring and increases the gas/metal area by 10-20 times the values of typical of top blowing. Rate of decarburization is higher than that of top blowing.

As in top blowing the intensity of decarburization at higher concentration of carbon is determined by the rate of supply of the oxidant. With a carbon content of less than 0.1% in bath, mass transfers become rate controlling which favours in iron oxidation. Since bottom blowing provides favourable conditions for decarburization reaction, the metal and slag have a lower degree of oxidation. For example, concentration of FeO in slag is not less than 15-25% during the entire blow in top blown steel-making and increases even more by the end of blowing, whereas in the bottom blown converter process it does not exceed 5-7% and raises up to 15-25%at low concentration of carbon. Since the slag in bottom blown converter has a low degree of oxidation at most during the whole operation, condition are unfavourable for dephosphorization reaction. Only toward the end phosphorous removal occurs.

Since the slag in the bottom blown converter oxidizes much less than in the top blowing, concentration of Mn in steel is higher in the course of a heat.

The advantages of pure top and pure bottom are coupled and a new technology is developed under the name combined top blowing of oxygen and bottom stirring. The advantages of such a technology when compared with pure top blowing like reduced FeO content of slag and O content of steel etc. are obvious.

Several process technologies for combined blowing were developed under different names. These technologies differ in

- Amount of inert gas
- Type, number and arrangement of bottom tuyeres and porous plugs.

Characteristic Feature of Converter Steelmaking

Supply of oxygen in the form of free gas jet is an important feature of converter steelmaking both in pure and different versions of combined blown ones. In this form of oxygen supply, the total time of blowing of oxygen is almost independent of converter capacity, oxygen blowing rate and bottom stirring. This was reflected by evaluating dimensionless momentum flow rate vs. ratio of time of blowing $(t)/\text{total}$ blowing time (t_{tot}) for different converter capacities ranging from 40-400 tons. Dimensionless momentum flow rate was correlated as

$$\frac{\dot{m}}{\rho\,lg \times L^{s}} = 7.25\times10^{-3}(\phi)^{0.78}$$

Where $(\phi) = {}^{t}\!/_{\text{total}}$

Dimensionless momentum flow rate describes the action of free oxygen jet produced by constant volume flow rate of oxygen at various lance distances. The dimensionless momentum flow rate number increases with the decrease in lance distance. Decrease in lance distance makes the blow hard and increase in lance distance makes the blow soft. Lance profile can be considered to generate soft blow initially and progressively harder blow with the progress of the blow.

The fundamental requirements of the lance profile in all converter steelmaking are formation of FeO rich slag in the initial stage and then removal of carbon and phosphorus by progressively increasing the availability of oxygen in the bath to avoid over oxidation of slag. The first requirement is achieved by “soft blow” (shallow penetration of jet) and the other requirement is achieved by hardening the blow (deep penetration of jet into bath) progressively. Thus soft and hard blow are essential requirement of refining of hot metal by impinging oxygen jet irrespective of the converter capacity and type of converter steelmaking practices (pure top blowing combined blowing) as a result the total oxygen blow time remains more or less same.

Environmental Issues in Oxygen Steelmaking

- Control of emissions during transfer of hot metal to desulphurization station.
- Disposition of slag
- Capture and removal of contaminants in hot and dirty exit gas from the converter.
- Particulate matter exiting with the exit gas.
- Emission of CO. for this purpose sufficient excess air must be used at the hood to burn CO.

Causes of High Turnover Rates of BOF

- Energetic supply of oxygen: This method ensures.
- Availability of oxygen where it is needed during refining
- Faster mechanism of mass transfer by producing droplets and slag/metal emulsion.
- Bottom stirring
- A basic and limy slag of required basicity is formed at the early stages of the blow.

Basic Oxygen Steelmaking

Oxygen converter being charged at ThyssenKrupp steel mill in Duisburg

Basic oxygen steelmaking (BOS, BOP, BOF, and OSM), also known as Linz–Donawitz-steelmaking or the oxygen converter process is a method of primary steelmaking in which carbon-rich molten pig iron is made into steel. Blowing oxygen through molten pig iron lowers the carbon content of the alloy and changes it into low-carbon steel. The process is known as *basic* because fluxes of burnt lime or dolomite, which are chemical bases, are added to promote the removal of impurities and protect the lining of the converter.

The process was developed in 1948 by Robert Durrer and commercialized in 1952–1953 by Austrian VOEST and ÖAMG. The LD converter, named after the Austrian towns Linz and Donawitz (a district of Leoben) is a refined version of the Bessemer converter where blowing of air is replaced with blowing oxygen. It reduced capital cost of the plants, time of smelting, and increased labor productivity. Between 1920 and 2000, labor requirements in the industry decreased by a factor of 1,000, from more than 3 man-hours per metric ton to just 0.003. The vast majority of steel manufactured in the world is produced using the basic oxygen furnace; in 2000, it accounted for 60% of global steel output. Modern furnaces will take a charge of iron of up to 400 tons and convert it into steel in less than 40 minutes, compared to 10–12 hours in an open hearth furnace.

History

The basic oxygen process developed outside of traditional "big steel" environment. It was developed and refined by a single man, Swiss engineer Robert Durrer, and commercialized by two small steel companies in allied-occupied Austria, which had not yet recovered from the destruction of World War II.

In 1856, Henry Bessemer patented a steelmaking process involving oxygen blowing for decarbonizing molten iron (UK Patent No. 2207). For nearly a hundred years commercial quantities of oxygen were not available at all or were too expensive, and the invention remained unused. During World War II German (C. V. Schwartz), Belgian (John Miles) and Swiss (Durrer and Heinrich Heilbrugge) engineers proposed their versions of oxygen-blown steelmaking, but only Durrer and Heilbrugge brought it to mass-scale production.

In 1943, Durrer, formerly a professor at the Berlin Institute of Technology, returned to Switzerland and accepted a seat on the board of Roll AG, the country's largest steel mill. In 1947 he purchased the first small 2.5-ton experimental converter from the U. S., and on April 3, 1948 the new converter produced its first steel. The new process could conveniently process large amounts of scrap metal with only a small proportion of primary metal necessary. In the summer of 1948, Roll AG and two Austrian state-owned companies, VOEST and ÖAMG, agreed to commercialize the Durrer process.

By June 1949, VOEST developed an adaptation of Durrer's process, known as the LD (Linz-Donawitz) process. In December 1949, VOEST and ÖAMG committed to building their first 30-ton oxygen converters. They were put into operation in November 1952 (VOEST in Linz) and May 1953 (ÖAMG, Donawitz) and temporarily became the leading edge of the world's steelmaking, causing a surge in steel-related research. Thirty-four thousand businesspeople and engineers visited the VOEST converter by 1963. The LD process reduced processing time and capital costs per ton of steel, contributing to the competitive advantage of Austrian steel. VOEST eventually acquired the rights to market the new technology. However, errors made by the VOEST and the ÖAMG management in licensing their technology made control over its adoption in Japan impossible and by the end of the 1950s the Austrians lost their competitive edge.

The original LD process consisted in blowing oxygen over the top of the molten iron through the water-cooled nozzle of a vertical lance. In the 1960s, steelmakers introduced bottom-blown converters and introduced inert gas blowing for stirring the molten metal and removing the phosphorus impurities.

In the Soviet Union, some experimental production of steel using the process was done in 1934, but industrial use was hampered by lack of efficient technology to produce liquid oxygen. In 1939, the Russian physicist Pyotr Kapitsa perfected the design of the centrifugal turboexpander. The process was put to use in 1942–1944. Most turboexpanders in industrial use since then have been based on Kapitsa's design and centrifugal turboexpanders have taken over almost 100 percent of the industrial gas liquefaction and in particular the production of liquid oxygen for steelmaking.

The big American steelmakers caught up late with the new technology; the first oxygen converters in the United States were launched at the end of 1954 by McLouth Steel in Trenton, Michigan, which accounted for less than 1 per cent of the national steel market. U.S. Steel and Bethlehem Steel introduced the oxygen process only in 1964. By 1970 half of the world's and 80% of Japan's steel output was produced in oxygen converters. In the last quarter of the 20th century use of basic oxygen converters for steel production was gradually partially replaced by the electric arc furnace using scrap steel/iron. In Japan the share of LD process decreased from 80% in 1970 to 70% in 2000; worldwide share of the basic oxygen process stabilized at 60%.

Process

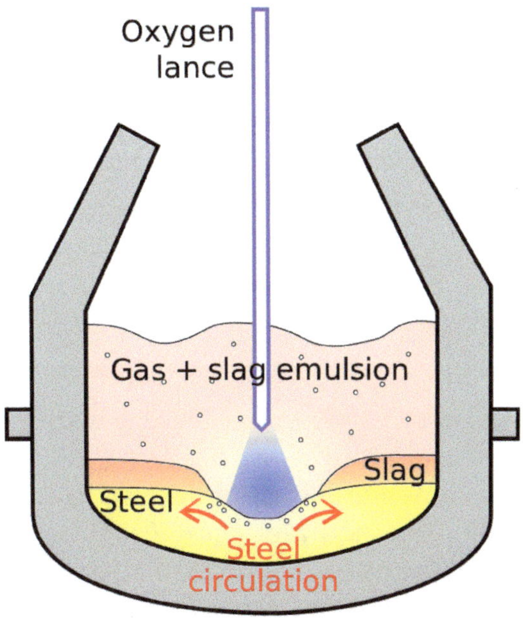

Principle of a LD converter

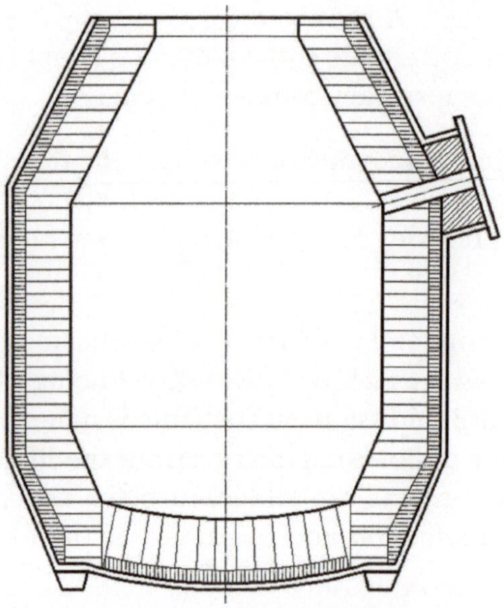

Cross-section of a basic oxygen furnace

The outside of a basic oxygen steelmaking plant at the Scunthorpe steel works.

Basic oxygen steelmaking is a primary steelmaking process for converting the molten pig iron into steel by blowing oxygen through a lance over the molten pig iron inside the converter. Exothermic heat is generated by the oxidation reactions during blowing.

The basic oxygen steel-making process is as follows:

1. Molten pig iron (sometimes referred to as “hot metal”) from a blast furnace is poured into a large refractory-lined container called a ladle;

2. The metal in the ladle is sent directly for basic oxygen steelmaking or to a pretreatment stage. High purity oxygen at a pressure of 700–1,000 kilopascals (100–150 psi) is introduced at supersonic speed onto the surface of the iron bath through a water-cooled lance, which is suspended in the vessel and kept a few feet above the bath. Pretreatment of the blast furnace hot metal is done externally to reduce sulphur, silicon, and phosphorus before charging the hot metal into the converter. In external desulphurising pretreatment, a lance is lowered into the molten iron in the ladle and several hundred kilograms of pow-

dered magnesium are added and the sulphur impurities are reduced to magnesium sulphide in a violent exothermic reaction. The sulfide is then raked off. Similar pretreatments are possible for external desiliconisation and external dephosphorisation using mill scale (iron oxide) and lime as fluxes. The decision to pretreat depends on the quality of the hot metal and the required final quality of the steel.

3. Filling the furnace with the ingredients is called *charging*. The BOS process is autogenous, i.e. the required thermal energy is produced during the oxidation process. Maintaining the proper *charge balance*, the ratio of hot metal, from melt, to cold scrap, is therefore very important. The BOS vessel can be tilted up to 360° and is tilted towards the deslagging side for charging scrap and hot metal. The BOS vessel is charged with steel or iron scrap (25–30%) if required. Molten iron from the ladle is added as required for the charge balance. A typical chemistry of hotmetal charged into the BOS vessel is: 4% C, 0.2–0.8% Si, 0.08%–0.18% P, and 0.01–0.04% S, all of which can be oxidised by the supplied oxygen except sulphur (which requires reducing conditions).

4. The vessel is then set upright and a water-cooled, copper tipped lance with 3–7 nozzles is lowered down into it and high purity oxygen is delivered at supersonic speeds. The lance "blows" 99% pure oxygen over the hot metal, igniting the carbon dissolved in the steel, to form carbon monoxide and carbon dioxide, and causing the temperature to rise to about 1700 °C. This melts the scrap, lowers the carbon content of the molten iron and helps remove unwanted chemical elements. It is this use of pure oxygen instead of air that improves upon the Bessemer process, as the nitrogen (a particularly undesirable element) and other gases in air do not react with the charge.

5. Fluxes (burnt lime or dolomite) are fed into the vessel to form slag, to maintain basicity above 3 and absorb impurities during the steelmaking process. During "blowing", churning of metal and fluxes in the vessel forms an emulsion, that facilitates the refining process. Near the end of the blowing cycle, which takes about 20 minutes, the temperature is measured and samples are taken. A typical chemistry of the blown metal is 0.3–0.9% C, 0.05–0.1% Mn, 0.001–0.003% Si, 0.01–0.03% S and 0.005–0.03% P.

6. The BOF vessel is tilted towards the slagging side and the steel is poured through a tap hole into a steel ladle with basic refractory lining. This process is called *tapping* the steel. The steel is further refined in the ladle furnace, by adding alloying materials to give it special properties required by the customer. Sometimes argon or nitrogen is bubbled into the ladle to make the alloys mix correctly.

7. After the steel is poured off from the BOS vessel, the slag is poured into the slag pots through the BOS vessel mouth and dumped.

Variants

Earlier converters, with a false bottom that can be detached and repaired, are still in use. Modern converters have a fixed bottom with plugs for argon purging. The Energy Optimization Furnace is a BOF variant associated with a scrap preheater where the sensible heat in the off-gas is used for preheating scrap, located above the furnace roof.

The lance used for blowing has undergone changes. Slagless lances, with a long tapering copper tip, have been employed to avoid jamming of the lance during blowing. Post-combustion lance tips burn the CO generated during blowing into CO_2 and provide additional heat. For slag-free tapping, darts, refractory balls and slag detectors are employed. Modern converters are fully automated with auto blowing patterns and sophisticated control systems.

Availability of Oxygen

Oxygen is available energetically during the refining. The energetic availability of oxygen is obtained by passing a certain flow rate of oxygen through the nozzle as shown in figure.

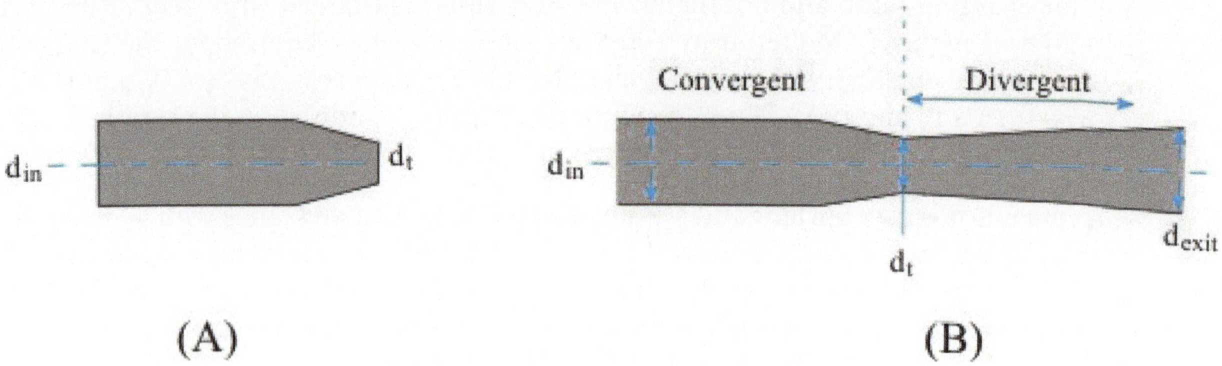

(A):Convergent nozzle, (B) Laval nozzle

In BOF steelmaking oxygen is blown through a laval nozzle. A Laval nozzle also called a convergent-divergent nozzle and is characterized by a flow passage whose cross sectional area decreases in the direction of flow and attains a minimum cross section area and then increases further in the direction of flow as shown in the figure (B). The minimum cross section area of the flow passage is called throat of the nozzle. The laval nozzle can accelerate the gas to the supersonic velocity.

Velocity of gas can be expressed in terms of Mach number

$$M = \frac{\text{velocity of gas}}{\text{velocity of sound}} = \frac{v}{a}$$

Thus at M= 1 gas exits the nozzle with the sonic velocity. Sonic velocity can be attained at the throat of the nozzle.

$M<1$ gas exits the nozzle at subsonic velocity.

$M>1$ gas exits the nozzle at supersonic velocity.

Convergent nozzle can accelerate gas at $M \leq 1$.

Convergent – Divergent nozzle can accelerate gas at $M \leq 1$.

In converter steelmaking Laval nozzles are used. Gas velocity at the exit corresponds to $M \approx 2 \text{ to } 2.4$ flow rate of oxygen is $3\,Nm^3/\text{min.ton}$ at pressures 10-14 bar. Multi –hole nozzles are used. Number of nozzles varies from 3 to 6.

Behaviour of Free Gas Jet

First we will see the behaviour gas when it exits a single Laval nozzle in the surrounding which consists of air. A gas on exiting through a nozzle spreads in the surrounding and is called "free gas jet", because spreading is not confined. Figure show a free jet in the surrounding. It is characterized by the potential (L_P) and supersonic (L_C) core length as shown in the figure.

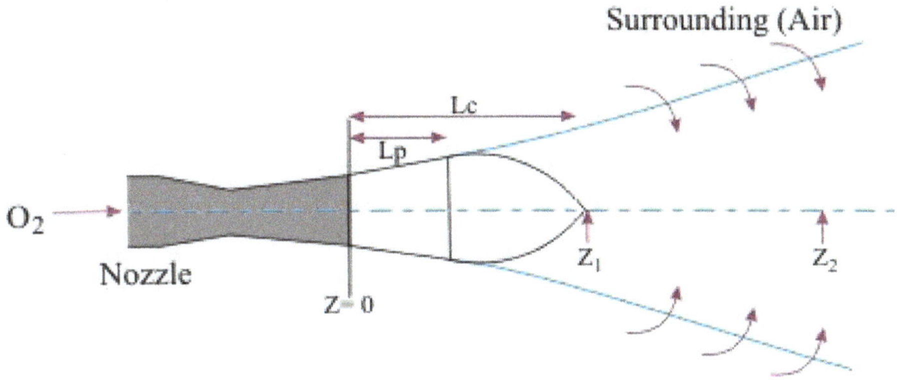

Discharge of pressurized fluid stream through a Laval nozzle

In the potential core no entrainment of the surrounding occurs and hence velocity of gas in both axial and radial direction is that at the exit value. Beyond the potential core both radial and axial velocity begins to decrease due to entertainment of the surrounding. Radial spreading of the jet can be seen in the figure. However a point Z_1 is reached in the free gas jet at which the gas velocity attains a sonic value (M=1). Within the length L_C gas velocity is above the supersonic value in both radial and axial direction. Beyond the supersonic core length the gas velocity is subsonic. Thus radial spreading and axial velocity decay beyond the potential core are the main characteristics of a free gas jet.

Due to spreading mass of the jet increases which means that concentration of the gas at plane Z=0 decreases due to entrainment of the surrounding. If oxygen is flowing through the nozzle, concentration of oxygen at plane Z_2 is lower than at Z_1 and at Z=0. But mass of jet (jet consists of main fluid + surrounding) at $Z_1 <$ mass at Z_2. Axial velocity of the jet is a function of axial distance measured from the nozzle exit.

One of the important property of the free jet is that it carries with it momentum flow rate which on hitting the liquid is converted into force and penetrates into the liquid. Momentum flow rate within the jet is conserved. This is an important property of the jet since it depends only on the upstream variables like pressure, number and diameter of the nozzle. It does not depend on the downstream conditions.

Behaviour of jets produced by multi-nozzles depends on number of nozzles and inclination angle (a) of each nozzle with the axis of the lance. Number of nozzles in converter steelmaking varies with the converter capacity but in general lay between 3 to 6. The inclination angle of each nozzle for a three hole, is $10-12°$ and for five holes is 15 to $16°$ with the axis of the lance.

Multi hole nozzles are compared with a single. The relationship between diameter of a single nozzle and the corresponding multi nozzles for the same flow rate of gas can be obtained from the area consideration in equation.

$$A_1 = N \times An$$

From equation 1 it follows that

$$d_1 = \sqrt{N} \times d_n$$

where d_1 is the diameter of a single nozzle and d_n is the diameter of each nozzle in a multi-nozzle and N is the number of nozzles.

The multi-free gas jets downstream the nozzle can coalesce or not would depend on inclination angle and number of nozzles for a given upstream pressure and flow rate of gas. A coalescing jet is similar to that of a single jet.

When angle of inclination is $10-12^\circ$ for a three hole lance, the multi-jets do not coalesce upto certain distance downstream the nozzle. A non-coalescing jet, when impinged on the liquid will produce penetration equal to number of jets.

The axial velocity decay and radial spreading depend on the ratio

$$\frac{\rho_{surr}}{\rho_j} = \frac{\text{density of surrounding}}{\text{density of jet}}$$

If $\frac{\rho_{surr}}{\rho_j} < 1$ i.e. $\rho_j > \rho_{surr}$ jet is denser than the surrounding. Such a jet will spread slowly in the surrounding; how slow would depend on the value of $\frac{\rho_{surr}}{\rho_j}$. Accordingly jet velocity will decay slower at any distance downstream the surrounding. This situation could be, for example a cold jet is discharged in the hot surrounding. Length of the potential core, L_P and length of the supersonic core L_C will be longer than at $\frac{\rho_{surr}}{\rho_j} = 1$.

If $\frac{\rho_{surr}}{\rho_J} > 1$ jet is lighter than surrounding; jet spreads faster which results in lower L_P and L_C than

$\frac{\rho_{surr}}{\rho_j} = 1$, for example cold jet is discharged into slag.

Action of Free Oxygen Jet

Velocity of the free oxygen jet is important. Axial velocity decreases as the distance downstream the nozzle increases due to entrainment of the surrounding. In the converter as the blow begins, the surrounding of oxygen jet is hot atmosphere. As the blow continues the jet surrounding changes from carbon monoxide to slag. For most of the periods the jet is submerged into slag. The surrounding in the converter is dynamic. The velocity of the jet depends on upstream pressure, downstream axial distance and the surrounding. It is difficult to calculate the jet velocity when the surrounding is changing, but the momentum flow rate within the get is independent of the distance downstream the nozzle and can be calculated from:

$$\dot{m} = 1.029 \times 10^5 P_O N d_n^2 \left[\left\{ 1 - \left(\frac{P}{P_O} \right)^{0.286} \right\} \right]^{0.5}$$

Where $\dot{m}$ = momentum flow rate in (Newton)

P_O = upstream stagnation pressure (bar)

N = Number of nozzles

P_O = Surrounding pressure

d_n = Throat diameter of the nozzle(m)

Upstream stagnation pressure can be calculated

$$P_O(\text{bar}) = 6.755(T)^{0.104}$$

T= capacity in tons.

Throat diameter is function of converter capacity and number of nozzles

$$d_n = \frac{7.46}{\sqrt{N \times 10}} T^{0.446}$$

For a 150 ton converter, diameter of each nozzle of a hole lance is calculated to be 0.035 m by equation . P_O is calculated to be 11.37 bar by equation and total momentum flow rate produced by a four hole nozzle is 4057 N by equation.

Jet carries with it momentum flow rate which on hitting the bath is converted into force. Thus action of free jet can be described in terms of dimensionless flow rate number (φ)

$$\varphi = \frac{\dot{m}}{\rho l g x^3}$$

ρ_l density of liquid (Kg/m^3), g (m/s^2) and X (m) is distance between the lance tip and liquid bath surface. If X is 3m and 1.5 m then $\varphi = 2.19 \times 10^{-3}$ and $= 0.0175$. We note that the dimensionless number increases with the decrease in the lance distance. Thus dimensionless flow rate number can be used to describe the dynamic variation of the lance distance. The dimensionless momentum flow rate number signifies the action of the gas jet on the bath at a distance X against the gravity.

Jet Penetrability

We calculate now the depth of penetration when the jet hits the bath surface at a distance x. Depth of penetration of an impinging jet (h) is

$$h(m) = 4.407 \times (\varphi)^{0.66}$$

At x = 3m, $h = 0.23$m and increases to 0.46m when the distance x is decreased to 1.5m. This means that there will be shallow jet penetration in the bath at x=3 m, whereas at x=1.5m jet will penetrate deep into the bath.

Dimensionless flow rate number describes the effect of lance distance on the penetrability of jet.

Shallow jet penetration as obtained at higher lance distance is a “soft jet” as compared to deep penetrating jet as obtained at lower distance and is termed “hard jet”.

This would mean that a constant volume flow rate of oxygen supplied at constant pressure when discharged through a nozzle can be made to hit the bath “soft “and can be made progressively harder.

Thus method of oxygen supply in converter steelmaking practice i.e. through “free jet” is very effective in terms of physico- chemical reactions.

The effects induced by a reactive soft and hard jet impinging oxygen jet, when it hits the hot metal bath are given below.

Soft Jet	Hard Jet
• Oxidation of Fe	• O_2 available deep in bath
• Shallow penetration	• C oxidation is favoured. P removal is impaired
• Slag/metal reaction	• CO evolution occurs deep into bath and its escape through the bath agitates the bath
• Slag formation is promoted. P removal is enhanced	• Droplets are produced which are then emulsified in the slag
Too long duration of soft let will promote sloping of slag due to overoxidation	

Process Control BOF Steelmaking

Steelmaking in BOF is very fast. It takes approximately 15 to 20 minutes for oxygen blowing and 50 to 60 minutes to tap molten steel. Liquid steel at turndown results from several non linear interconnected complex processes like gas/ liquid metal reaction as dependent on oxygen availability, slag/metal reaction as dependent on physic-chemical properties of slag and faster reaction rate induced by three phase dispersion (CO+ slag +metal droplets).

Control of the above processes is needed so that for a given input of hot metal, scrap, flux and oxygen flow rate, steel of the desired chemistry is produced with minimum loss of iron in slag.

How are the Above Processes Controlled?

Rates of gas/metal, slag/metal and gas/slag/metal droplet reaction are controlled by lance profile, oxygen flow rate and bottom stirring rate, it is required to raise or lower the lance distance for a given oxygen flow rate and bottom stirring rate so that steel of desired composition and tapping temperature can be obtained within the stipulated blowing time. For this purpose we need to develop process models which can describe the process quantitatively. These models must be supported by the data for accurate predictions for the future requirements.

Process Control Models

One of the objectives of the process control models is to predict turndown composition and temperature of steel so that unnecessary blowing of oxygen is not required. For a long time human expertise was the control tool. The operator used to deliver instructions to exercise the process control. These controls were human specific.

Developments in computer has resulted into development of sophisticated models, like

a) Static model

b) Semi-dynamic model

c) Dynamic model

Static Model

Static models are based on materials and heat balance by considering initial and final states of reactants. In the material balance, mass of all input and output elements is considered. Once mass balance is done then heat balance is done.

Sensible heat of all inputs+ Heat produced or absorbed by oxidation reaction= Heat taken out by steel, slag and exit gases and fumes+ heat losses from the converter mouth and through the lining of the converter+ any other heat losses.

In making materials and heat balance some assumptions may be required for example

- iron loss in slag,
- carbon removal in the form of CO,
- complete dissolution of CaO in slag
- basicity of slag
- thermo-physical and thermo-chemical properties of slag and metal

The above are some of the assumptions, further may be added.

By Coupling of Mass and Balance One Can Predict

i) Quantity of hot metal and scrap

ii) Amount of flux

iii) Total quantity of oxygen required to be blown.

iv) Amount of slag produced

v) Volume of exit gases

It is very much important that the prediction based upon the model is verified by the actual plant data. Tuning of the model is necessary because the predictions are based on equilibrium considerations and uncertainties due to simplified assumptions.

Reliability of predictions increases when the predictions of the model are compared with the plant data for large number of heats. Statistical correlations can be developed and used to fine tune the model. For this purpose it is of utmost necessary to collect the reliable data from the reliable instruments.

BOF steelmaking is a stochastic process. Oxygen blowing produces lot of turbulent in the phases during hot metal refining. The amount of droplets emulsified in the slag, amount of lime dissolution, carbon removal rate, intensity of oxygen jet impinging the bath, lance distance, bottom stirring rate due to plugging of tuyeres etc. may vary from one heat to the other. The error in predictions may be due to:

i) Error in weighing

ii) Differences in lime dissolution from one heat to the other

iii) Effect of size of scrap on its dissolution. Large size will take more time to dissolve as compared with smaller ones.

iv) State of foaming of slag and entrained metal droplets in slag. This may vary from one heat to other due to behavior of oxygen jet in a dynamic surrounding. Surrounding of oxygen jet changes during the blow. As a result extent and magnitude of slag/metal reactions might change.

v) Converter lining profile due to wear.

vi) Extent of mixing within the phases and between the phases.

Semi – dynamic Model

The above features make static control models inadequate. Hence improvements are required. Static model predictions can be improved by measurements on temperature and composition of slag and metal by a sub lance during the blow.

Immersion type sensors can also be used to measure carbon and oxygen concentration during the blow. All these data are collected and fed into the computer which compares the model predictions to suggest the action to be taken by the operator. These are semi-dynamic model. It must be noted that error in carbon measurement may need to over blow the heat when end carbon measured is greater than specified carbon. Also if end point carbon measured is lower than the specified, carburization has to be done in the ladle for carbon adjustment.

Determination of carbon by a sub lance is indirectly done by measuring the temperature of steel through a sample collected by the sub lance during the flow.

Dynamic Models

Static models do not calculate the variation of blowing parameters as a function of time. For this purpose continuous measurement of some representative parameter is required during the blow.

In a preprogrammed model, these values are fed continuously and corrections can be done during the blow. Dynamic models contain all features of static model; in addition it includes reaction kinetics and process dynamics. In steelmaking major reactions are oxidation of carbon and iron. In a dynamic model rate of decarburization and O_2 consumed are determined from exit analysis. Computation of Fe is then done.

Dynamic control requires measurements continuously. Exit gas temperature and its composition can be measured continuously and the data can be fed to computer. From the exit gas composition and temperature carbon, oxygen in bath can be determined as a function of time. From carbon balance, decarburization rate can be determined. Oxygen balance provides the following information:

- Total rate of oxidation
- Relative fraction of oxygen reacting with carbon and iron and other elements

Enthalpy balance based on exit gas temperature and composition and amount gives information on energy leaving the system.

Slag height can also be determined by measuring the acoustic sound produced by the slag during the blow. Sonic meter measures the intensity of sound. Intensity of sound is fed to computer which in turn adjusts the lance height, oxygen flow rate and bottom stirring rate to control the slag/metal height.

Modern Trends in BOF Steelmaking

Post Combustion

The oxidation of carbon to CO in BOF steelmaking is the principle reaction. In converter steelmaking CO produced during refining exits the furnace in exhaust system where it is combusted with the ambient air.

Combustion of CO to CO_2 produces large amount of thermal energy

$$CO+\frac{1}{2}O_2=CO_2 \qquad -\Delta H_f^o=283\times10^3\,\frac{kJ}{kg\,mole}$$

$$=12634\times10^3\,\frac{kJ}{m^3}$$

Thus post combustion of CO in BOF and transferring the heat of combustion to the slag and metal offers an additional amount of energy. The amount of post combustion taking place in the furnace can be represented by post combustion ratio (P CR)

$$PCR=\frac{\%CO_2}{\%CO+CO_2}$$

Technology of Post Combustion

A technology is required which can inject oxygen in the converter just above the slag so that CO can be combusted to CO_2. The supply of oxygen for the post combustion must be well distributed above the slag surface for an efficient combustion of CO. Transfer of the heat to the slag and metal phase is also to be considered.

One possible way is to add several small orifices around the main supersonic nozzle tip. Lance distance has to be adjusted so that oxygen through the orifice is available for combustion of CO to CO_2 above the slag surface.

Still another method could be a lance with double flow for oxygen. In the main oxygen lance, a separate oxygen inlet and oxygen control system can be provided which is solely dedicated to post combustion. This design may provide better control of oxygen for post combustion without affecting the oxygen flow through the main lance.

One of main requirement of oxygen flow for post combustion is that velocity and angle of oxygen flow should be low to avoid the refractory wear. Nozzle diameter, angle of oxygen flow and location of nozzle are the principle design issues.

Potential Post Combustion Issues

- Repair costs of post combustion lance are higher.
- Post combustion lances are not rigid as standard oxygen lances.
- Optimization of angle, location, number and diameter of ports for supply of oxygen needs to be established keeping in mind the refractory wear.
- Development of high quality refractory would be required because high temperature would be generated in the post combustion zone.

Slag Splashing

Splashing of slag to coat the refractory lining has become a standard practice to increase the lining life. Slag splashing is done as follows:

- At the end of BOS process, steel is drained of and slag is retained in the vessel.
- The O_2 lance is lowered and high pressure N_2 is used to splash molten slag on the walls of BOS vessels for a period of 2 to 4 min.
- Slag refractory provides a consumable refractory lining which protects the furnace lining.
- The excess slag is poured out.

What is required for slag splashing?

- Compositional adjustment of end slag against FeO and MgO concentration. Presence of MgO in the end slag should be greater than 13% to produce a high temperature phase and to increase slag viscosity.

- FeO reduces melting temperature of slag and increases the amount of low melting phase in slag. Low melting phase ensures good adhesion between slag and refractory and high melting phase $MgO.Fe_2O_3$ imparts erosion resistance to the lining.
- Al_2O_3 content of slag should be low.
- For $\frac{CaO}{SiO_2}$ ratio greater then 5, SiO_2 should be added.

Figure shows schematic diagram of slag splashing in a converter

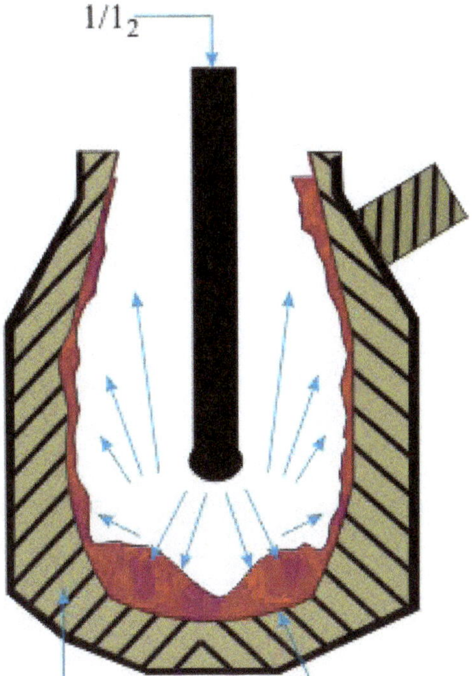

Schematic representation of slag splashing in converter

Liquidus Temperature of splashing slag:

The liquidus temperature of slag is important because

i) Slag layer should contain enough high temperature phases to prevent attack by slagduring BOF process.

ii) Low melting phase melts and thereby coating thickness decreases

iii) Too high liquidus temperature will lead to non uniform distribution of slag and there is tendency for the slag to build up at the bottom.

Problems with Slag Splashing

A. Lance skulling

This is particularly severe if some metal is left in the vessel

B. Vessel mouth skulling.

C. Blockage of tuyeres.

D. Sensors are embedded in the refractories to detect the presence of slag during tapping. The slag coating interferes with the detection.

Benefits of Slag Splashing

- Longer life of furnace lining (over 60000 heats).
- More recycling of BOF slags.
- Less CaO is needed in BOF process due to dissolution of basic slag coating.
- Decrease in sloping results in increase in yield.
- Rapid formation of slag occurs due to melting of low temperature phase of the coating.

Slag Free Tapping

Priamary steelmaking slag contains FeO, SiO_2, P_2O_5 and CaO. FeO content varies form 15-18 %. These slags are oxidizing in nature. During tapping, carry-over of slag should be avoided for the following reasons:

- During ladle treatment impurities can revert back from slag to metal.
- During synthetic slag practice in ladle, composition of slag will be altered which will affect secondary steelmaking refining operations like desulphurization, deoxidation etc. This will lead to increased consumption of deoxidizers.
- Carried-over slag can increase the refractory wear.
- FeO and MnO of slag react with Al and forms Al_2O_3 which is a solid inclusion at the steel-making temperature, this will affect steel cleanliness.

Mechanism of Slag Carry-over

Slag can be carried-over from converter to ladle in the initial stage when converter is tilted. As tapping proceeds converter is tilted further which leads to formation of vortex and top slag is carried into the ladle. During up-tilting, part of the slag may be carried into ladle.

The amount of drained slag depends on tilting speed of converter, bath level, drain velocity.

Prevention of Slag Carry-over

Several methods are developed and in use to minimize slag carry-over. Tap-hole plug, slag cut ball pneumatic slag stopper and slag detection methods are commonly is use.

Electric Furnace Steelmaking

Steelmaking in electric arc furnace has emerged as an important steelmaking process in recent years .The flexibility and easy adoptability of EAF steelmaking to accommodate the fluctuating market demand have evolved into the concept of mini steel plants to produce different grades of finished products (long or flat or mixed) of plain carbon or alloy steels from scrap and other metallic charge materials. Although scrap is the preferred raw material but sponge iron and iron carbide are being used regularly in most plants because of shortage of steel scrap and to dilute the concentration of tramp elements.Several developments in the design and operation have made EAF steelmaking to contribute significantly to the overall total production of steel in the world. According to an estimate, the proportion of electric steel is around 40 to 45% in the total world steel production. It must be noted that EAF consumes lot of electric energy and hence the cost and availability of electrical power are important issues in electric steel development.

Type of Electric Furnaces

In principle an electric arc is formed between the electrode and the metallic charge and charge is heated from the arc radiation. Electric arc furnaces are of two type (a) alternating current and (b) direct current. In alternating current, furnace operates by means of electric current flowing from one electrode of three to another through the metallic charge. In direct current, the current flows from carbon electrode, which acts as cathode, to an anode embedded in the bottom of the furnace.

Construction of AC Electric Arc Furnace

The furnace consists of a steel shell, lined with suitable refractory materials and is mounted on the tilting mechanism. The shell thickness is around 0.005 times the shell diameter. Three electrodes enter through the roof. The hood may be swung away for charging. Heat is generated by the hot area formed between the electrodes and the charge.

Hearth

The hearth contains metal and slag. The hearth lining consists of backing lining and working lining. The backing lining is few layers of high fired magnesite bricks on which working lining is rammed with either dolomite or magnesite mass. Permeable blocks or porous refractory elements are introduced through the bottom to inject inert gas for stirring. The EAF steel bath is shallow; the aspect ratio of the bath is around 0.2 to 0.22.

Roof

The roof is exposed to more heat than other furnace elements. Its lining is also subjected to radiant heat reflected from the walls and slag. High alumina bricks and magnesite – chromite bricks are used for roof lining. The roof lining is water cooled which increases the life of refractory lining to at least 10-20 times more than without water cooling. The roof has three holes to allow insertion of the electrodes.

Electrode

A typical alternating current operated EAF has three electrodes. Electrodes are round in section, and typically in segments with threaded coupling, so that as the electrodes wear, new segments can be added. Graphite electrodes are preferred over carbon electrodes because of better electrical conductivity. The electrodes are automatically raised and lowered by a positioning system.

Electrode Consumption Depends On

- Oxidation of the surface of the electrode
- Mechanical losses due to fracture
- Dissolution in slag during carbon boil

The diameter of the electrode should correspond to the current supplied; if current density is excessively high, electrodes will be heated and oxidized vigorously. The electrode current could vary from 12 to $16A / cm^2$ for 400 to 600 m electrode diameter. Larger electrode diameter increases electric energy consumption.

The electrodes are positioned at apexes of an equilateral triangle. The diameter of the circle passing through the centers of electrodes is called the diameter of the electrode spacing. If the electrodes are placed close to each other and far from furnace walls, the charge at the furnace banks will be heated belatedly. With large spacing diameter, electric arcs will burn near the walls, which will result in rapid wear of the lining.

The electrode spacing diameter for the bath diameter could be 0.45 for small furnaces, 0.35 for medium- sized and large furnaces, and still lower for super- powerful furnaces. For a bath diameter of 5560 mm of a 100 ton furnace the electrode spacing diameter would be $0.35 \times 5560 = 1946mm$.

Side Walls

The side walls refractory materials should be able to withstand thermal shock and corrosive action of slag. Hot spot is formed on the side walls due to the radiation from arc flames, reflected from bath surface during power input. The side wall is lined with magnesite, dolomite or chrome magnesite bricks up to the slag line. The side wall thickness is usually 450 to 500mm for 10 to 50 ton furnaces and 550 to 650mm for 100 to 200 ton furnaces.

Transformer Power:

Electric furnaces are powerful consumers of electric energy. The operating voltage of a furnace is 100-800V and the current may reach several thousand amperes. The furnace transformer transforms high voltage energy into low voltage. The melting process consists of two periods: meltdown and refining period. In melt down period higher electric energy is required as compared with the refining period.

In small furnaces, the power consumption for melting is about $600k\,Wh/ton$ and it falls to $450k\,Wh/ton$ in big furnaces. Additional 150 to 400 $k\,Wh/ton$ power is required during refining.

Depending on the Practice

Large transformers are required to run electric arc furnaces. During melting more power is required than during refining. The transformer capacity is designed to suit melting requirements. The capacity of the transformer is usually 470- 650 KVA per tonne of furnace capacity. In terms of hearth area, the transformer capacity is in the range of 750-900 KVA per square meter.

Charging Materials:

Steel scrap is the principle raw material. It may constitute 60 to 80% of the charge. In some practices sponge iron and or pig iron is also used for chemical balance. In basic furnaces slag formers like limestone, fluorspar, sand, and quartzite are used to form a slag to refine the metal. For decarburization oxygen lancing is used. Iron ore is also added. Ferro-manganese, ferrosilicon or aluminium are used for deoxidation. To produce alloy steels, alloying elements are added.

Plant Layout

Layout of an electric arc furnace steelmaking shop varies from plant to plant due to difference in the quality of the product and the scale of production. Some plants have just one EAF while others have two. The variation is also due to whether the shop is provided with oxygen lancing and carbon injection facilities, gas cleaning equipments and finished castings or ingots. Broadly electric furnace steelmaking shop comprises of the following:

1. Electric furnace
2. Transport facilities for ladle
3. Scrap charging
4. Auxiliary injection facilities
5. Electrode movement mechanism
6. Charging of raw materials and weighing system
7. Slag disposal.

In an ideal layout, all the above facilities should be arranged so as to ensure smooth input and output of materials.

Arc Furnaces Operation

It consists of charging, melt down period and refining. The large baskets containing heavy and light scrap are preheated through the exit gas. Burnt lime and spar are added to help early slag formation. Iron one or mill scale may also be added if refining is required during melt- down period.

The roof is swung off the furnace, and the furnace is charged. Some furnaces are equipped with continuous charging. Hot metal is also charged as per the requirement.

In the meltdown period, electrodes are lowered and bored into the scrap. Lower voltages are selected in order to protect the roof and walls from excessive heat and damage from the arcs. Once the arc is shielded by scrap, voltage is increased to form molten metal pool to reduce the meltdown period. During meltdown period, silicon, manganese and carbon oxidizes. Also oxidizing and limy slag is produces which promotes dephosphorization as well. Melt- down time depends on

- Arc conditions: larger arc requires lower current and lower heat losses
- Deep or shallow bath: deep bath shortens the meltdown period.

Refining continues even during melting. Removal of phosphorus must be complete before the rise in temperature and carbon boil.

The single oxidizing slag practice is employed when removal of sulphur is not required. When both P and S are required to be removed double slag practice is used. In double slag practice, oxidizing slag is removed and reducing slag is formed after deoxidation with ferrosilicon or ferromanganese or aluminum. Reducing slag helps to avoid loss of alloying elements.

Once the bath chemistry and its temperature are attained, heat is deoxidized and finished for tapping.

Comparison with Oxygen Steelmaking

	EAF	Oxygen steelmaking
Source of energy	Electric + chemical energy	Chemical energy; Autogeneous process
Iron–containing raw material	Hot metal + directly reduced iron + scrap in the suitable proportion as per practice	Hot metal + 20 – 30% scrap
Operating procedure	Oxygen lancing is to promote decarburization, scrap melting and post combustion.	Oxygen supply is continuously done to refine hot metal to steel. A three phase dispersion of slag/metal/gas forms to accelerate the refining rates.
	Slag foaming is induced to shield refractory lining from the heat of arc.	
	Carbon injection is done to induce foamy slag practice	

The growth of electric steel production around the world has been driven by lower investment, higher operational flexibility and easy adoptability to market demand on long or flat products of either plain carbon or alloy steels. Growth has been supported by updating installations and technologies to reduce the electric energy, electrode consumption and tap to tap time. Figure shows the developments in electric steelmaking technologies.

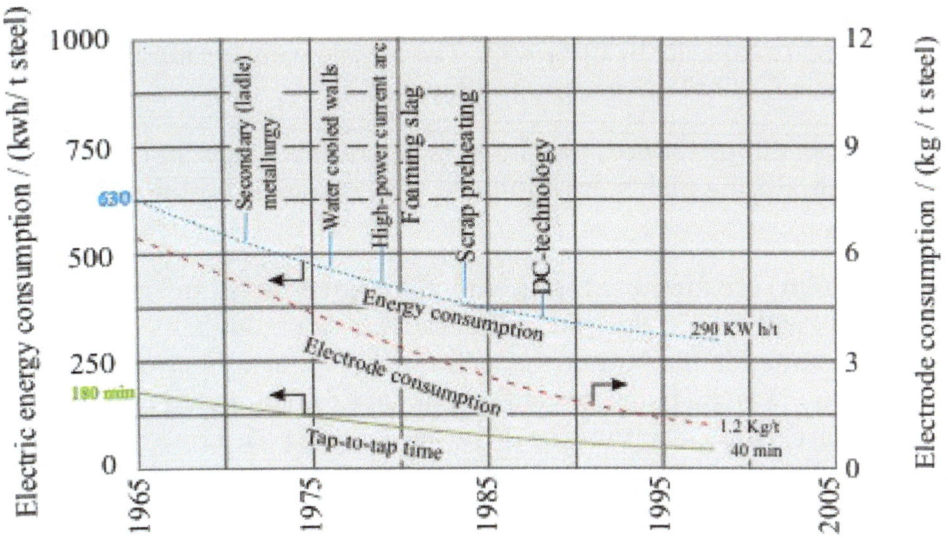

Trends in EAF steelmaking technologies developments in EAF steelmaking

Developments in EAF technologies are strongly supported by secondary steelmaking. One can note in the figure that the power consumption has decreased from 630 Kwh/ton of steel to 290kWh/ton. Similarly tap tp tap time has decreased from 180 minutes to 40 minutes and electrode consumption has decreased from arounf 6.2 kg/ton to as low as 1.2 kg/ton within the periods of representation in the figure. This became possible with the several simultaneous developments in the secondary steel-making method. Table shows the various developments.

Developments in EAF Steelmaking

Furnace design	Process operating technologies	Chemical energy	Charge materials
Split shell design	Bottom stirring	Oxidation reaction	Directly reduced iron, hot metal
Transformer power	Foamy slag practice	Post combustion oxy fuel burner	
DC arc furnace	Scrap preheating	Carbon injection	

Furnace design:

1. Construction of hearth and lower side section of the shell of larger diameter than the top opening. This leads to increase in shell volume which results in larger tonnage charge, lower heat losses and improved thermal efficiency.
2. In the split shell design, shell structure is constructed in two sections: lower section which contains hearth and free board allowance for slag, and upper section containing side wall and roof. The two sections are coupled such that the upper section can be repaired easily. This reduces the downtime and increases furnace availability.
3. High powered transformers are the current trends. Most modern furnaces operate at 500k VA/ton and the trend is towards ultra high power ranging in between 700k VA/ton to 1000k VA/ton. Developments are in progress to install transformer with 1500k VA/ton capacity. It is claimed that a 120 tons operating at 180 MVA transformer capacity and by

using refining combined burner technology through oxygen gas and carbon injection, it is possible to increase capacity by up to 50%. The largest transformer in AC EAF corresponds to a rated power of 240 MVA for 300 ton furnace.

4. Eccentric bottom taping reduces tap times, temperature losses and slag carry over into ladle. The strip producing plants are equipped with eccentric bottom tapping in electric arc furnaces.

5. DC (direct current) arc furnaces represent a different concept in arc furnace design. Most DC furnaces are with single electrode where current flows down from the carbon electrode to an anode mounted in the bottom of the furnace. Reduced electrode consumption of the order of 50 to 60 % is the major benefit of a dc furnace compared to a convectional three-phase arc furnace. Noise levels for the dc furnaces are lower. Lower maintenance costs are claimed and refractory costs are less for sidewall but more for the furnace bottom. A dc arc furnace requires an addition of the bottom electrode (anode), a dc reactor, and a thyristor all of which add cost to a dc furnace. The electrode technology limits diameter to a maximum of 700 mm allowing a dc current of 100kA and 70 MVA power for single electrode furnace. Furnace size is limited to 200 tons. Further developments are in progress.

Process Operating Technologies

Most of the developments in process operating technologies are in AC- electric arc furnaces as these furnaces are popular.

i) Bottom stirring

In convectional arc furnaces there is little natural electrical turbulence within the bath. Due to absence of stirring large piece of scrap can take a long time to melt and may require oxygen lancing. Argon or nitrogen stirring

- Eliminates temperature and concentration gradients
- Shortens tap-to-tap times
- Reduces refractory, electrode and power consumption and
- Improves yield of iron and alloys

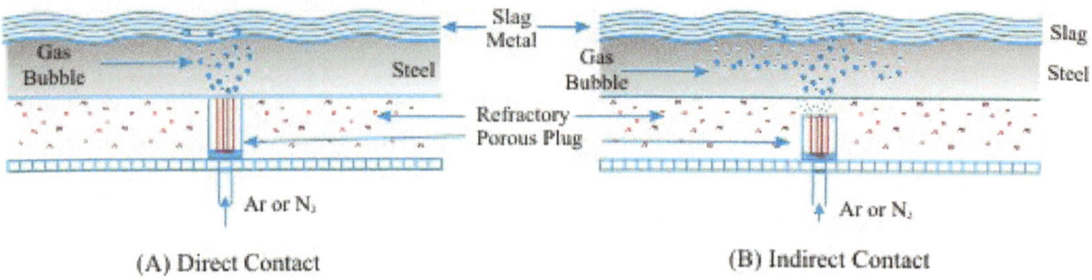

Industrial bottom stirring systems in electric arc furnace

Industrial systems for bottom stirring are either with direct contact plug or with indirect contact plug. In direct contact plug, the plug is in contact with molten metal, whereas in indirect one the plug is embedded in a porous bottom refractory. In the indirect contact, the plug is not directly in contact the molten metal. The gas enters the bath via the porous refractory hearth which results in stirring over a large area when compared with direct plug as shown in the figure. It shows the direct contact and indirect contact plug for bottom stirring. Note that in indirect contact large area of the bath is stirred as compared with direct contact plug.

ii) Foamy slag practice

In EAF steelmaking, progressive melting of scrap increases the irradiative heat transfer from arc to the side walls of the furnace. By covering the arc in a layer of slag, the arc is shielded and more energy is transferred to the bath. The foaming slag during this period is beneficial. The effectiveness of slag foaming depends on slag basicity, FeO content of slag, slag temperature and availability of carbon to react with either oxygen or FeO of slag.

Slag foams in steelmaking due to entrapment of gas bubbles. Gas producing reactions in steelmaking are:

a) Reaction between FeO of slag with carbon

$$(\mathrm{FeO})_1 + \mathrm{C} = [\mathrm{Fe}] + \{\mathrm{CO}\}$$

b) Between carbon and oxygen dissolved in metal

$$[\mathrm{C}] + [\mathrm{CO}] = \{\mathrm{CO}\}$$

c) Between chromium oxide and carbon:

$$\mathrm{Cr_2O_3} + 3\mathrm{C} = 2\mathrm{Cr} + 3\mathrm{CO}$$

Reactions above are important in carbon steelmaking whereas the last reaction is important in stainless steel making.

Injection of carbon and oxygen at several places in the bath assures slag foaming practice, when carbon content of the bath is insufficient. Typically carbon injection rates for slag foaming are 2.5 to 5 kg/ ton of steel. In high powered furnaces carbon injection is 5-10 kg/ton of steel.

In stainless steel making Cr_2O_3 forms in preference toFeO due to higher affinity of Cr to oxygen. The solubility of chromium oxide in the slag is considerably weaker in comparison to FeO for the same basicity and thermal conditions. The oxygen/carbon injection technique in the high chromium alloy steel production and to foam the slag is difficult. Moreover, additional chromium will also be lost due to oxygen injection.

The novel technology utilizes the reduction of iron and chromium oxide by carbon as well as thermal dissociation of limestone contained in small briquette.

iii) Scrap preheating

Preheating of scrap brings thermal energy into the furnace. Preheating of scrap to $540^\circ\mathrm{C}$ brings $81\,\mathrm{kwh/ton}$ of additional energy. Scrap preheating gives the following advantages:

i) Reduction in energy consumption by 40-60 kwh/ton depending on the scrap preheat temperature

ii) Electrode consumption reduces by 0.3 to 0.36 kg/ton

iii) Refractory consumption decreases by 0.9 to 1.4 kg/ton

iv) Tap to tap time reduces by 5 to 8 minutes.

It is important to note that scrap preheating technology needs to be developed. Thermal energy is required to preheat the scrap and is economical only when the waste heat from the furnace is utilized. For this purpose it is important to know the energy balance of the electric furnace. The energy balance of an EAF, as shown in the figure indicates that 20% of the total energy leaves the furnace in the waste gases and represents about 130 kWh/ton of steel produced. Efficient utilization of thermal energy of exit gas is the key to realize the advantages of preheating of scrap. Batch preheating and continuous preheating are the available technologies. In CONSTEEL, scrap and exit gases move counter current to each other. It is possible to preheat the scrap to $\sim 320°C$.

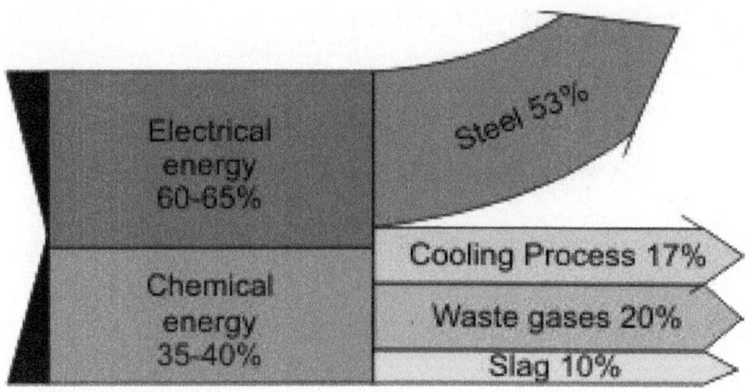

Energy balance of an electric furnace steelmaking

Usage of Chemical Energy

The high electrical energy costs pushed EAF steelmakers to look for alternative energy sources. One such source is the chemical energy derived from chemical reactions. In recent years the chemical energy supply amounts to 35% to 40% of the total energy in most of the modern EAFs.

Oxidation Reactions

The main oxidation reactions are oxidation of iron and carbon besides oxidation of silicon and manganese. The oxidation of iron though generates more energy than oxidation of carbon but iron oxidation results in loss in productivity.

$$Fe+0.5O_2=FeO;\ \text{Heat contant } 6kW/m^3O_2$$

$$C+0.5O_2=CO;\ \text{Heat contant } 3.5kW/m^3O_2$$

Hence oxygen injection must be controlled such that iron oxidation is kept minimum. For bath carbon levels above 0.3%, all oxygen reacts with carbon to produce CO. Below 0.3%, the efficiency of carbon oxidation to form CO drops and more and more FeO is generated in the slag.

For scrap carbon levels below 0.1%, FeO levels in the slag can be quite high and represents an unavoidable yield loss. Increased carbon injection is necessary to control slag FeO levels and to prevent excessive refractory wear. Efficiency of heat transfer from oxidation reactions is extremely high due to the fact that these reactions are occurring in the bath. Greater penetrability of oxygen jet ensures the occurrence of oxidation reactions in the bath.

i) Post combustion

It is a practice of generating additional energy for melting steel by using the right amount of extra oxygen to combust CO and H_2 which evolve within the EAF. Carbon monoxide is generated in an EAF by

- Partially combusted hydrocarbons entering the furnace with the scrap
- Combustion of charged and injected carbon via $C+0.5O_2=CO$
- The reduction of FeO by carbon during slag foaming

Hydrogen is generated by:

- The cracking of hydrocarbons (oil in scrap, methane)
- The reduction of water: $H_2O+CO=H_2+CO_2$ or $H_2O+C=H_2+CO$

In EAF, carbon monoxide and hydrogen may be available at the freeboard, whereas the foaming slag contains carbon- monoxide. It must be noted that oxygen flow should have low velocity to promote mixing with the furnace gases and to avoid scrap oxidation and rebound of oxygen from the scrap to the water cooled panels. Post combustion in the slag typically aims at combustion of 20 to 30% of the CO generated in slag and 70 – 80% at the free board.

For post combustion speed of oxygen injection must be low and also uniform distribution of oxygen is required.

ii) Oxy – fuel burner

Oxy-fuel burner uses natural gas or oil, together with pure oxygen to produce an extremely high flame temperature. Oxy- fuel burners are used to melt unmelted scrap between the electrodes and to provide heat to cold spot. On most modern UHP furnaces, the primary function of burners is to provide heat to cold spots to ensure even scrap melting and to decrease the melting time necessary to reach a flat bath. Typically industry practice indicates that 0.133 MW of burner rating should be supplied per ton of furnace capacity. Others recommend $32\,\text{kWh/ton}$ of burner power to eliminate cold spots in a UHP furnace and 50 to $200\,\text{kWh/ton}$ of burner power for low powered furnaces.

iii) Carbon injection

Injection of carbon brings following benefits:

1. For 100 percent scrap practice or when carbon content of the bath is insufficient to produce CO for slag foaming, carbon injection is beneficial.
2. Carbon oxidation produces CO which on post combustion generates thermal energy.

It is to be noted that carbon injection requires oxygen injection to onset carbon oxidation.

Requirements for Chemical Energy Usage

The chemical energy usage requires to develop a device to inject oxygen in different modes:

1. Hold mode (to prevent plugging)
2. Burner mode (to heat and melt scrap)
3. Soft lancing mode (for post combustion)
4. Supersonic lancing mode (for decarburization and slag foaming)
5. Carbon injection mode (when slag foaming is required)

Injectors are either fixed type or moveable type. Submercible hand lances arc used through the slag door. Large opening in EAF shell is required. Slag and metal splashing restricts the device movement. Excessive repairs and down- time are associated with this technology.

An innovative design is CO-jet injectors which are fixed type and can be mounted on the furnace shell. CO-jet injectors are highly flexible in usage.

Future of EAF Steelmaking

The EAF has the largest growth potential both in terms of production capacity and technology evolution. Future EAF will be equipped with all modern technologies- like Ultra high power input (up to $1500 \mathrm{kVA/t}$, latest oxygen and carbon injection technology and design features- like ultra high shell design, heavy mill type components.

This combination leads to an Electric Arc Furnace where the tap to tap times can be extremely short and the corresponding productivity reaches the level of larger furnace sizes or converter plants.

The two main reasons for this are:

- The possibility of a higher electrical power input and
- A far higher efficiency of chemical energy, decarburization and scrap preheating compared to the same size (tap weight) standard furnace.

It is interesting to compare a conventional 120 ton EAF with the ultimate 120 ton EAF.

Conventional 120 ton EAF	**Ultimate 120 ton EAF**
2- bucket charge Scrap bucket 130 m^3 Furnace volume 145 m^3	1-bucket charge Scrap bucket 185 m^3 Furnace volume 210 m^3
Transformer design upto 1,000kVA/t, 120MVA for 120 ton tapping weight, Secondary voltage up to 1,200V	Transformer design upto 1,500kVA/t, 180MVA for 120 ton tapping weight, Secondary voltage up to 1,500V

Utilization of chemical energy 3 oxygen gas burners 3 refined combined burners (RCB) 2 carbon injectors	Utilization of chemical energy 3 oxygen gas burners 5 refined combined burners (RCB) 4 carbon injectors 4 post combustion injectors

Refined Combined Burner (RCB) technology combines a conventional oxy/gas burner with a supersonic oxygen injection lance and is designed to optimize the injection of carbon and oxygen into EAF. It supplies chemical energy through chemical reactions of fuel and gas, oxygen, and carbon injected into the furnace.

Alternative Charge Materials

Raw materials and operating practices affect EAF efficiency and yield. The traditional EAF charge has been 100% cold scrap. The charge materials saturation is critical for several reasons:

- The product mix served by EAFs is moving more towards value-added steels, which are specified with low metallic residuals and low nitrogen levels (automotive flat rolled, cold heading- rolled and wire).
- The availability of scrap needed to meet the requirements of value added products is limited. The availability of scrap is decreasing as more and more near-net-shape metalworking operations appear.
- Yield and energy consumption are both strongly dependent on the quality and physical characteristics of the charge materials availability.

Type of Metallic Charge Materials

- Pig iron/hot metal
- Direct reduced iron (DRI)
- Iron carbide (Fe_3C)

Pig Iron

Pig iron is an intermediate product of the iron industry. Pig iron has a very high carbon content, typically 3.5–4.5%, along with silica and other constituents of dross, which makes it very brittle, and not useful directly as a material except for limited applications. Pig iron is made by smelting iron ore into a transportable ingot of impure high carbon-content iron in a blast furnace as an ingredient for further processing steps. The traditional shape of the molds used for pig iron ingots was a branching structure formed in sand, with many individual ingots at right angles to a central channel or *runner*, resembling a litter of piglets being suckled by a sow. When the metal had cooled and hardened, the smaller ingots (the *pigs*) were simply broken from the runner (the *sow*), hence the name *pig iron*. As pig iron is intended for remelting, the uneven size of the ingots and the inclusion of small amounts of sand caused only insignificant problems considering the ease of casting and handling them.

Pig iron of a type used to make ductile iron, stored in a bin

History

Casting pig iron, Iroquois smelter, Chicago, between 1890 and 1901.

Smelting and producing pig iron and other iron products was known to the Ancient Egyptians and gradually spread around the Eastern Mediterranean as far as Ancient Greece. The Roman Empire and later the Muslim caliphates of the Middle Ages inherited and refined these technologies. Because of the collapse of the Western Roman Empire, Western Europe did not rediscover the process until the Late Middle Ages (1325–1500). The phase transition of the iron into liquid in the furnace was an *avoided* phenomenon, as decarburizing the pig iron into steel was an extremely tedious process using medieval technology.

The Chinese were also making pig iron by the later Zhou Dynasty (which ended in 256 BC).

Uses

Traditionally pig iron was worked into wrought iron in finery forges, later puddling furnaces, and more recently into steel. In these processes, pig iron is melted and a strong current of air is directed over it while it is stirred or agitated. This causes the dissolved impurities (such as silicon) to be thoroughly oxidized. An intermediate product of puddling is known as *refined pig iron, finers metal*, or *refined iron*.

Pig iron can also be used to produce gray iron. This is achieved by remelting pig iron, often along with substantial quantities of steel and scrap iron, removing undesirable contaminants, adding alloys, and adjusting the carbon content. Some pig iron grades are suitable for producing ductile iron. These are high purity pig irons and depending on the grade of ductile iron being produced these pig irons may be low in the elements silicon, manganese, sulfur and phosphorus. These types of pig irons are used to dilute all the elements (except carbon) in a ductile iron charge which may be harmful to the ductile iron process.

Modern Uses

Until recently, pig iron/slag was typically poured directly out of the bottom of the blast furnace through a trough into a ladle car for transfer to the steel mill in mostly liquid form; in this state, the pig iron was referred to as *hot metal*. The hot metal was then poured into a steelmaking vessel to produce steel, typically an electric arc furnace, induction furnace or basic oxygen furnace, where the excess carbon is burned off and the alloy composition controlled. Earlier processes for this included the finery forge, the puddling furnace, the Bessemer process, and the open hearth furnace.

Modern steel mills and direct-reduction iron plants transfer the molten iron to a ladle for immediate use in the steel making furnaces or cast it into pigs on a pig-casting machine for reuse or resale. Modern pig casting machines produce stick pigs, which break into smaller 4–10 kg piglets at discharge.

Direct Reduced Iron

Direct-reduced iron (DRI), also called sponge iron, is produced from the direct reduction of iron ore (in the form of lumps, pellets or fines) to iron by a reducing gas or elementary carbon produced from natural gas or coal. Many ores are suitable for direct reduction.

Reduced iron derives its name from the chemical change that iron ore undergoes when it is heated in a furnace at high temperatures in the presence of hydrocarbon-rich gases, carbon monoxide or elementary carbon. Direct reduction refers to processes which reduce iron oxides to metallic iron at temperatures below the melting point of iron. The product of such solid state processes are called direct reduced iron. The reducing gas is a mixture of gases, primarily hydrogen (H_2) and carbon monoxide (CO). The process temperature is typically 800 to 1200 °C.

Process

Direct reduction processes can be divided roughly into two categories, gas-based, and coal-based. In both cases, the objective of the process is to drive off the oxygen contained in various forms of iron ore (sized ore, concentrates, pellets, mill scale, furnace dust etc.), in order to convert the ore, without melting it (below 1200 °C), to metallic iron.

The direct reduction process is comparatively energy efficient. Steel made using DRI requires significantly less fuel, in that a traditional blast furnace is not needed. DRI is most commonly made into steel using electric arc furnaces to take advantage of the heat produced by the DRI product.

In modern times, direct reduction processes have been developed to specifically overcome the difficulties of conventional blast furnaces. DRI is successfully manufactured in various parts of the world. The initial investment and operating costs of direct reduction plants are low compared to integrated steel plants and are more suitable for developing countries where supplies of coking coal are limited.

Factors that help make DRI economical:

- Direct-reduced iron has about the same iron content as pig iron, typically 90–94% total iron (depending on the quality of the raw ore) as opposed to about 93% for molten pig iron, so it is an excellent feedstock for the electric furnaces used by mini mills, allowing them to use lower grades of scrap for the rest of the charge or to produce higher grades of steel.
- Hot-briquetted iron (HBI) is a compacted form of DRI designed for ease of shipping, handling, and storage.
- Hot direct reduced iron (HDRI) is iron not cooled before discharge from the reduction furnace, that is immediately transported to a waiting electric arc furnace and charged, thereby saving energy.
- The direct reduction process uses pelletized iron ore or natural "lump" ore. One exception is the fluidized bed process which requires sized iron ore particles.
- The direct reduction process can use natural gas contaminated with inert gases, avoiding the need to remove these gases for other use. However, any inert gas contamination of the reducing gas lowers the effect (quality) of that gas stream and the thermal efficiency of the process.
- Supplies of powdered ore and raw natural gas are both available in areas such as Northern Australia, avoiding transport costs for the gas. In most cases the DRI plant is located near natural gas source as it is more cost effective to ship the ore rather than the gas.
- this method produces 97% pure iron.

India is the world's largest producer of direct-reduced iron, a vital constituent of the steel industry. Many other countries use variants of the process, so providing iron for local engineering industries.

Problems

Directly reduced iron is highly susceptible to oxidation and rusting if left unprotected, and is normally quickly processed further to steel. The bulk iron can also catch fire since it is pyrophoric. Unlike blast furnace pig iron, which is almost pure metal, DRI contains some siliceous gangue, which needs to be removed in the steel-making process.

History

Producing sponge iron and then working it was the earliest method used to obtain iron in the Middle East, Egypt, and Europe, where it remained in use until at least the 16th century. There is some evidence that the bloomery method was also used in China, but China had developed blast furnaces to obtain pig iron by 500 BCE.

The advantage of the bloomery technique is that iron can be obtained at a lower furnace temperature, only about 1,100 °C or so. The disadvantage, relatively to using a blast furnace, is that only small quantities can be made at a time.

Uses

Sponge iron is not useful by itself, but can be processed to create wrought iron. The sponge is removed from the furnace, called a *bloomery*, and repeatedly beaten with heavy hammers and folded over to remove the slag, oxidise any carbon or carbide and weld the iron together. This treatment usually creates wrought iron with about three percent slag and a fraction of a percent of other impurities. Further treatment may add controlled amounts of carbon, allowing various kinds of heat treatment (e.g. "steeling").

Today, sponge iron is created by reducing iron ore without melting it. This makes for an energy-efficient feedstock for specialty steel manufacturers which used to rely upon scrap metal.

Pig iron is a good charge material because of its

- High density
- Low melting point
- Carbon contribution
- No tramp elements

It must be known that EAF is designed exclusively to use electrical energy. Pig iron contains $3-4\%C, 0.1-0.2\%P, 0.6-0.8\%Mn$ and $0.6-0.8\%Si$. Charging of pig iron means refining is to be done in EAF. For example if we make a charge mix containing 20% hot metal with 3.5%C and 80% scrap with 0.1%C to produce 100 ton steel with 0.1% carbon. We need to remove around 640kgC. To remove this amount of carbon we would be needing $\approx 630\,m^3$ oxygen on the assumption that oxygen utilization is 100% and is used for removal of carbon only. Under actual conditions we would be needing more oxygen to remove carbon if we take into account oxidation of $Fe \rightarrow FeO, Si \rightarrow SiO_2, Mn \rightarrow MnO\ and\ P \rightarrow P_2O_5$. More proportion of pig iron in charge mix will increase oxygen requirement. What we learn from this simple calculation is that one has to supply oxygen under EAF condition and has to handle large amount of CO as well. Extra decarburization will increase tap to tap time and may decrease productivity. Under practical condition 30% hot metal is suggested to give optimum results with regard to productivity, electric consumption etc. The source of hot metal is either blast furnace or smelting reduction process. Hot metal can only be used when EAF plant is in close proximity of the blast furnace or smelting reduction units, otherwise one has to use pig iron and extra energy would be required to melt the pig iron.

Direct Reduced Iron

Direct reduced iron (DRI) or hot briquetted iron (HBI) has emerged as an important substitute of the scrap. DRI contain small amount(0.02%)of tramp element; hence substituting DRI for scrap leads to dilution of tramp elements in steel.

DRI is produced by reduction of iron ore with carbon or gaseous reducing agent. The product of gaseous reductant is HBI.

DRI contains: Free iron +oxygen combined with iron+ free carbon+ gangue minerals (SiO_2, Al_2O_3, P_2O_5 etc).

Quality of DRI is important. Quality of DRI can be judged in terms of metallization and oxygen remaining. Metallization (M) is defined as

$$M = \frac{\text{Free Iron} \times 100}{\text{Total Iron}}$$

Higher is the metallization, less oxygen would be required to remove oxygen from iron oxide of DRI during steelmaking.

Consider 1000kg iron ore of composition $80\% Fe_2O_3$ and 20% gangue minerals. Reduction of iron ore produces DRI in which oxygen is present as FeO.

Metallization (%)	Free iron (kg)	Fe-FeO (kg)	Gangue (kg)	Oxygen with FeO(kg)
80	448	112	200	32
85	476	84	200	24
90	504	56	200	16
95	532	28	200	08

We note from the table the following:

a) Increase in metallization increases free iron in DRI which is good.

b) Increase in metallization decreases iron which is combined with FeO. Lower amount of heat and reductant would be required to recover iron of FeO from DRI. Reduction of FeO with C is endothermic. Energy is required to compensate for the endothermic reaction.

c) Increase in metallization decreases the gangue minerals in DRI. Gangue minerals would be recovered as slag; decrease in gangue minerals will decrease slag volume during production of steel using high metalized DRI as a charge material.

d) Because of b) and c), the overall effect would be decrease in electric consumption and reduced electrode wear.

The benefits arising due to the use of high metalized DRI in charge materials must be considered along with the production cost of metalized burden. Overall economics of metalized DRI in electric steelmaking is important while deciding % metallization in the feed.

Another important issue in case of DRI is the proportion of DRI in the charge. For this purpose 1000 kg sponge iron of 90% metallization is to be used in different proportions. We consider charge mix is scrap and DRI. Calculations show that 1288.66 kg iron ore would be needed to produce 1000 Kg DRI with 90% metallization. The following table illustrates the influence of proportion of DRI on the free iron and other variables of DRI:

DRI (%)	Free iron (kg)	Fe-FeO(kg)	Gangue (kg)
10	65	7	26
30	195	22	77
40	260	29	103
50	325	36	129

We note the following:

i) Increase in proportion of DRI increases the free iron in the charge.

ii) Fe which is combined with oxygen as FeO increases with increase in the proportion of DRI. This would require extra amount of heat energy both chemical and thermal to reduce FeO to Fe and to raise the temperature of reactants to $1500-1600^{o}C$.

iii) Gangue content of the DRI increases with the increase in proportion of DRI. This would lead to increase in slag volume and heat load.

What is the Consequence?

A) Tap to tap time may decrease with the increase in DRI but more proportion of DRI may increase tap to tap time for two reasons, namely more time is required to reduce $FeO \rightarrow Fe$ and, to handle increase volume of slag.

B) Increased proportion of DRI beyond a limit may increase electrode consumption and refractory wear.

Carbon Content in DRI

Carbon content in DRI helps reducing FeO to Fe. Excess carbon than required to reduce FeO would require oxygen injection. Oxygen in DRI is $0.286 \times FeO$. Reduction of FeO is

$$FeO + C = Fe + CO.$$

According to reaction above carbon requirement is $0.17\, kg/kg\, FeO$.Carbon content in excess of $0.17\, kg/kg\, FeO$ would require oxygen injection and CO will be generated. Excess carbon aids chemical energy which reduces electrical energy requirement. While producing DRI from carbononaceous reductant, excess carbon than reduction for FeO requirement is to be maintained. In DRI produced from gaseous reductant, extra carbon can be mixed with DRI or otherwise carbon injection is to be done.

The amount of carbon required (in kg) to reduce iron oxide (FeO) content of the sponge iron is as follows [@]:

$$C = 1.67\left[100 - \%M - \left(\frac{\%Sl}{100} \times \%Fe\right)\right]$$

Where M is the degree of metallization, Sl is the amount of slag, and Fe is the amount of iron in slag.

It must be known that DRI is non conductor of electricity. Hence hot heel (molten metal pool) is required to melt DRI. In procedure hot heel is created by melting scrap or keeping liquid metal of earlier heat. This method makes easy melting of DRIU.

Iron Carbide

Other charging material is iron carbide which contains 6% carbon. It is prepared by heating iron oxide fines in a mixture of CO, CO_2, methane, H_2 and water vapour at temperatures between 550.

Charging Methods for DRI

In small furnaces (lower than 5T) batch charging is preferred. Continuous charging brings advantages like

- Less power off-time.
- Heat losses resulting from delays are eliminated.
- Lower electrical losses.
- Reaction between carbon and FeO of DRI produces strong carbon boil during charging which improves heat transfer and slag/metal mixing.
- Charging and refining take place simultaneously which reduces tap to tap time.

Continuous charging is preferred when the sponge iron or DRI is around 60%; the limitation is due to the small furnace capacity. However, ultra high powered furnaces can operate with 100% sponge iron.

References

- Pigott, Vincent C. (1999). The Archaeometallurgy of the Asian Old World. Philadelphia: University of Pennsylvania Museum of Archaeology and Anthropology. ISBN 0-924171-34-0, p. 186-187
- P. King, 'The Cartel in Oregrounds Iron: Trading in the Raw Material for Steel During the 18th century", Journal of Industrial History 6 (2003), 25–48
- Wagner, Donald B. (2001). The State and the Iron Industry in Han China. Copenhagen: Nordic Institute of Asian Studies Publishing. ISBN 87-87062-83-6, pp. 80–83
- Samuel Thomas (September 1899). "Reminiscences of the Early Anthracite-Iron Industry". Transactions of The American Institute of Mining Engineers. Retrieved 5 December 2016
- Williams, Alan R. (2003). The Knight and the Blast Furnace: a History of the Metallurgy of Armor in the Middle Ages & the Early Modern Period. Leiden: Brill. ISBN 9789004124981, pp 883

4

Steel Mill and Ladle Metallurgy

The industrial plants that manufacture steel are called steel mills. They also design casting-products such as blooms, billets and slabs. A ladle in a foundry is a vessel that pours molten steel. Casting ladle, treatment ladle, and transfer ladle are three types of ladles based on their respective uses. The chapter closely examines the key concepts of steel mills and ladle metallurgy to provide an extensive understanding of the subject.

Steel Mill

Integrated steel mill in the Netherlands. The two large towers are blast furnaces.

A steel mill or steelworks is an industrial plant for the manufacture of steel.

Steel mill can refer to the steel works making rolled products from iron ore, but it also designs, more precisely the plant where steel semi-finished casting products (blooms, ingots, slabs, billets) are made, from molten pig iron or from scraps.

History

Since the invention of the Bessemer process, steel mills have replaced ironwork, based on puddling or fining methods. New ways to produce steel appeared later: from scraps melted in an electric arc furnace and, more recently, from direct reduced iron processes.

In the late 19th and early 20th centuries the world's largest steel mill was the Barrow Hematite Steel Company steelworks located in Barrow-in-Furness, United Kingdom. Today, the world's largest steel mill is in Gwangyang, South Korea.

Integrated Mill

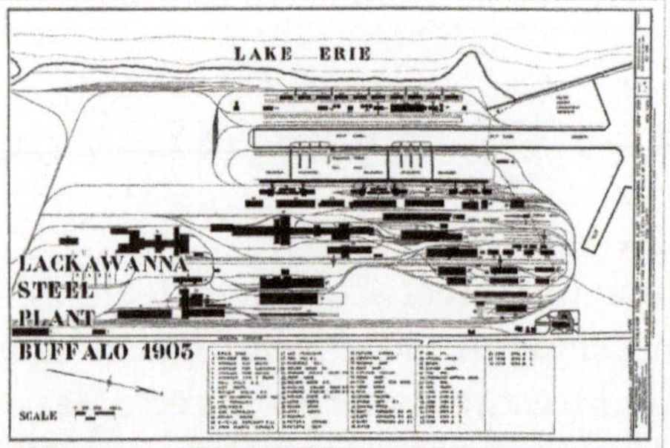

Plan of the Bethlehem Steel plant in Buffalo, New York ca. 1903, showing the various elements of an integrated steel mill

Blast furnaces of Třinec Iron and Steel Works

Interior of a Steel Mill

An integrated steel mill has all the functions for primary steel production:

- iron making (conversion of ore to liquid iron),
- steel making (conversion of pig iron to liquid steel),
- casting (solidification of the liquid steel),
- roughing rolling/billet rolling (reducing size of blocks)
- product rolling (finished shapes).

The principal raw materials for an integrated mill are iron ore, limestone, and coal (or coke). These materials are charged in batches into a blast furnace where the iron compounds in the ore give up excess oxygen and become liquid iron. At intervals of a few hours, the accumulated liquid iron is tapped from the blast furnace and either cast into pig iron or directed to other vessels for further steel making operations. Historically the Bessemer process was a major advancement in the production of economical steel, but it has now been entirely replaced by other processes such as the basic oxygen furnace.

Molten steel is cast into large blocks called *blooms*. During the casting process various methods are used, such as addition of aluminum, so that impurities in the steel float to the surface where they can be cut off the finished bloom.

Because of the energy cost and structural stress associated with heating and cooling a blast furnace, typically these primary steel making vessels will operate on a continuous production campaign of several years duration. Even during periods of low steel demand, it may not be feasible to let the blast furnace grow cold, though some adjustment of the production rate is possible.

Integrated mills are large facilities that are typically only economical to build in 2,000,000-ton per year annual capacity and up. Final products made by an integrated plant are usually large structural sections, heavy plate, strip, wire rod, railway rails, and occasionally long products such as bars and pipe.

A major environmental hazard associated with integrated steel mills is the pollution produced in the manufacture of coke, which is an essential intermediate product in the reduction of iron ore in a blast furnace.

Integrated mills may also adopt some of the processes used in mini-mills, such as arc furnaces and direct casting, to reduce production costs.

World integrated steel production capacity is at or close to world demand, so competition between suppliers results in only the most efficient producers remaining viable. However, due to the large employment of integrated plants, often governments will financially assist an obsolescent facility rather than take the risk of having thousands of workers thrown out of jobs.

Minimill

An ingot of steel entering a rolling mill

A minimill is traditionally a secondary steel producer; however, Nucor (one of the world's largest steel producers), as well as one of its competitors, Commercial Metals Company (CMC) use minimills exclusively. Usually it obtains most of its iron from scrap steel, recycled from used automobiles and equipment or byproducts of manufacturing. Direct reduced iron (DRI) is sometimes used with scrap, to help maintain desired chemistry of the steel, though usually DRI is too expensive to use as the primary raw steelmaking material. A typical mini-mill will have an electric arc furnace for scrap melting, a ladle furnace or vacuum furnace for precision control of chemistry, a strip or billet continuous caster for converting molten steel to solid form, a reheat furnace and a rolling mill.

Originally the mini mill was adapted to production of bar products only, such as concrete reinforcing bar, flats, angles, channels, pipe, and light rails. Since the late 1980s, successful introduction of the direct strip casting process has made mini mill production of strip feasible. Often a mini mill will be constructed in an area with no other steel production, to take advantage of local markets, resources, or lower-cost labour. Mini mill plants may specialize, for example, in making coils of rod for wire-drawing use, or pipe, or in special sections for transportation and agriculture.

Capacities of mini mills vary: some plants may make as much as 3,000,000 tons per year, a typical size is in the range 200,000 to 400,000 tons per year, and some old or specialty plants may make as little as 50,000 tons per year of finished product. Nucor Corporation, for example, annually produces around 9,100,000 tons of sheet steel from its four sheet mills, 6,700,000 tons of bar steel from its 10 bar mills and 2,100,000 tons of plate steel from its two plate mills.

Since the electric arc furnace can be easily started and stopped on a regular basis, mini mills can follow the market demand for their products easily, operating on 24-hour schedules when demand is high and cutting back production when sales are lower.

Foundry

From Fra Burmeister og Wain's Iron Foundry, by Peder Severin Krøyer, 1885.

A foundry is a factory that produces metal castings. Metals are cast into shapes by melting them into a liquid, pouring the metal in a mold, and removing the mold material or casting after the metal has solidified as it cools. The most common metals processed are aluminium and cast iron. However, other metals, such as bronze, brass, steel, magnesium, and zinc, are also used to produce castings in foundries. In this process, parts of desired shapes and sizes can be formed.

Process

A Foundryman, pictured by Daniel A. Wehrschmidt in 1899.

In metalworking, casting involves pouring liquid metal into a mold, which contains a hollow cavity of the desired shape, and then allowing it to cool and solidify. The solidified part is also known as a casting, which is ejected or broken out of the mold to complete the process. Casting is most often used for making complex shapes that would be difficult or uneconomical to make by other methods.

Melting

Melting metal in a crucible for casting

A metal die casting robot in an industrial foundry

Melting is performed in a furnace. Virgin material, external scrap, internal scrap, and alloying elements are used to charge the furnace. Virgin material refers to commercially pure forms of the primary metal used to form a particular alloy. Alloying elements are either pure forms of an alloying element, like electrolytic nickel, or alloys of limited composition, such as ferroalloys or master alloys. External scrap is material from other forming processes such as punching, forging, or machining. Internal scrap consists of gates, risers, defective castings, and other extraneous metal oddments produced within the facility.

The process includes melting the charge, refining the melt, adjusting the melt chemistry and tapping into a transport vessel. Refining is done to remove deleterious gases and elements from the molten metal to avoid casting defects. Material is added during the melting process to bring the final chemistry within a specific range specified by industry and/or internal standards. Certain fluxes may be used to separate the metal from slag and/or dross and degassers are used to remove dissolved gas from metals that readily dissolve certain gasses. During the tap, final chemistry adjustments are made.

Furnace

Several specialised furnaces are used to heat the metal. Furnaces are refractory-lined vessels that contain the material to be melted and provide the energy to melt it. Modern furnace types include electric arc furnaces (EAF), induction furnaces, cupolas, reverberatory, and crucible furnaces. Furnace choice is dependent on the alloy system quantities produced. For ferrous materials EAFs, cupolas, and induction furnaces are commonly used. Reverberatory and crucible furnaces are common for producing aluminium, bronze, and brass castings.

Furnace design is a complex process, and the design can be optimized based on multiple factors. Furnaces in foundries can be any size, ranging from small ones used to melt precious metals to furnaces weighing several tons, designed to melt hundreds of pounds of scrap at one time. They are designed according to the type of metals that are to be melted. Furnaces must also be designed based on the fuel being used to produce the desired temperature. For low temperature melting point alloys, such as zinc or tin, melting furnaces may reach around 500°C. Electricity, propane, or natural gas are usually used to achieve these temperatures. For high melting point alloys such as steel or nickel-based alloys, the furnace must be designed for temperatures over 1600° C. The fuel used to reach these high temperatures can be electricity (as employed in electric arc furnaces) or coke.

The majority of foundries specialize in a particular metal and have furnaces dedicated to these metals. For example, an iron foundry (for cast iron) may use a cupola, induction furnace, or EAF, while a steel foundry will use an EAF or induction furnace. Bronze or brass foundries use crucible furnaces or induction furnaces. Most aluminium foundries use either electric resistance or gas heated crucible furnaces or reverberatory furnaces.

Degassing

Degassing is a process that may be required to reduce the amount of hydrogen present in a batch of molten metal. Gases can form in metal castings in one of two ways:

1. by physical entrapment during the casting process or
2. by chemical reaction in the cast material.

Hydrogen is a common contaminant for most cast metals. It forms as a result of material reactions or from water vapor or machine lubricants. If the hydrogen concentration in the melt is too high, the resulting casting will be porous; the hydrogen will exit the molten solution, leaving minuscule air pockets, as the metal cools and solidifies. Porosity often seriously deteriorates the mechanical properties of the metal.

An efficient way of removing hydrogen from the melt is to bubble a dry, insoluble gas through the melt by purging or agitation. When the bubbles go up in the melt, they catch the dissolved hydrogen and bring it to the surface. Chlorine, nitrogen, helium and argon are often used to degas non-ferrous metals. Carbon monoxide is typically used for iron and steel.

There are various types of equipment that can measure the presence of hydrogen. Alternatively, the presence of hydrogen can be measured by determining the density of a metal sample.

In cases where porosity still remains present after the degassing process, porosity sealing can be accomplished through a process called metal impregnating.

Mold Making

Diagrams of two Pattern Types

A diagram of draft on a pattern

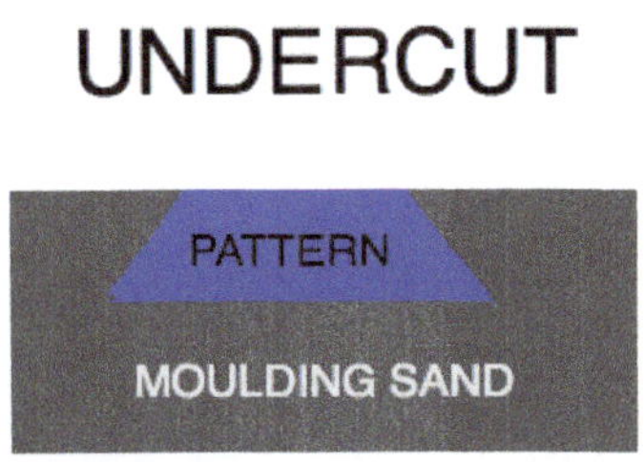

A diagram of an undercut in a mold

In the casting process a pattern is made in the shape of the desired part. Simple designs can be made in a single piece or solid pattern. More complex designs are made in two parts, called split patterns. A split pattern has a top or upper section, called a cope, and a bottom or lower section called a drag. Both solid and split patterns can have cores inserted to complete the final part shape. Cores are used to create hollow areas in the mold that would otherwise be impossible to achieve. Where the cope and drag separates is called the parting line.

When making a pattern it is best to taper the edges so that the pattern can be removed without breaking the mold. This is called draft. The opposite of draft is an undercut where there is part of the pattern under the mold material, making it impossible to remove the pattern without damaging the mold.

The pattern is made out of wax, wood, plastic, or metal. The molds are constructed by several different processes dependent upon the type of foundry, metal to be poured, quantity of parts to be produced, size of the casting, and complexity of the casting. These mold processes include:

- Sand casting — Green or resin bonded sand mold.
- Lost-foam casting — Polystyrene pattern with a mixture of ceramic and sand mold.
- Investment casting — Wax or similar sacrificial pattern with a ceramic mold.
- Ceramic mold casting — Plaster mold.
- V-process casting — Vacuum with thermoformed plastic to form sand molds. No moisture, clay or resin required.
- Die casting — Metal mold.
- Billet (ingot) casting — Simple mold for producing ingots of metal, normally for use in other foundries.

Pouring

Bronze poured from a crucible into a mold, using the lost-wax casting process

In a foundry, molten metal is poured into molds. Pouring can be accomplished with gravity, or it may be assisted with a vacuum or pressurized gas. Many modern foundries use robots or automatic pouring machines to pour molten metal. Traditionally, molds were poured by hand using ladles.

Shakeout

The solidified metal component is then removed from its mold. Where the mold is sand based, this can be done by shaking or tumbling. This frees the casting from the sand, which is still attached to the metal runners and gates — which are the channels through which the molten metal traveled to reach the component itself.

Degating

Degating is the removal of the heads, runners, gates, and risers from the casting. Runners, gates, and risers may be removed using cutting torches, bandsaws, or ceramic cutoff blades. For some metal types, and with some gating system designs, the sprue, runners, and gates can be removed by breaking them away from the casting with a sledge hammer or specially designed knockout machinery. Risers must usually be removed using a cutting method but some newer methods of riser removal use knockoff machinery with special designs incorporated into the riser neck geometry that allow the riser to break off at the right place.

The gating system required to produce castings in a mold yields leftover metal — including heads, risers, and sprue (sometimes collectively called sprue) — that can exceed 50% of the metal required to pour a full mold. Since this metal must be remelted as salvage, the yield of a particular gating configuration becomes an important economic consideration when designing various gating schemes, to minimize the cost of excess sprue, and thus overall melting costs.

Heat Treating

Heat treating is a group of industrial and metalworking processes used to alter the physical, and sometimes chemical, properties of a material. The most common application is metallurgical. Heat treatments are also used in the manufacture of many other materials, such as glass. Heat treatment involves the use of heating or chilling, normally to extreme temperatures, to achieve a desired result such as hardening or softening of a material. Heat treatment techniques include annealing, case hardening, precipitation strengthening, tempering, and quenching. It is noteworthy that while the term "heat treatment" applies only to processes where the heating and cooling are done for the specific purpose of altering properties intentionally, heating and cooling often occur incidentally during other manufacturing processes such as hot forming or welding.

Surface Cleaning

After degating and heat treating, sand or other molding media may remain adhered to the casting. To remove any mold remnants, the surface is cleaned using a blasting process. This means a granular media will be propelled against the surface of the casting to mechanically knock away the adhering sand. The media may be blown with compressed air, or may be hurled using a shot wheel. The cleaning media strikes the casting surface at high velocity to dislodge the mold remnants (for example, sand, slag) from the casting surface. Numerous materials may be used to clean cast surfaces, including steel, iron, other metal alloys, aluminium oxides, glass beads, walnut shells, baking powder, and many others. The blasting media is selected to develop the color and reflectance of the cast surface. Terms used to describe this process include cleaning, bead blasting, and sand blasting. Shot peening may be used to further work-harden and finish the surface.

Finishing

The final step in the process of casting usually involves grinding, sanding, or machining the component in order to achieve the desired dimensional accuracies, physical shape, and surface finish.

Modern foundry (circa 2000)

Removing the remaining gate material, called a gate stub, is usually done using a grinder or sander. These processes are used because their material removal rates are slow enough to control the amount of material being removed. These steps are done prior to any final machining.

After grinding, any surfaces that require tight dimensional control are machined. Many castings are machined in CNC milling centers. The reason for this is that these processes have better dimensional capability and repeatability than many casting processes. However, it is not uncommon today for castings to be used without machining.

A few foundries provide other services before shipping cast products to their customers. It is common to paint castings to prevent corrosion and improve visual appeal. Some foundries assemble castings into complete machines or sub-assemblies. Other foundries weld multiple castings or wrought metals together to form a finished product.

More and more, finishing processes are being performed by robotic machines, which eliminate the need for a human to physically grind or break parting lines, gating material, or feeders. Machines can reduce risk of injury to workers and lower costs for consumables — while also increasing productivity. They also limit the potential for human error and increase repeatability in the quality of grinding.

Ladle (Foundry)

A ladle of molten iron is poured into an open hearth furnace for conversion into steel at Allegheny Ludlum Steel Corp., 1941

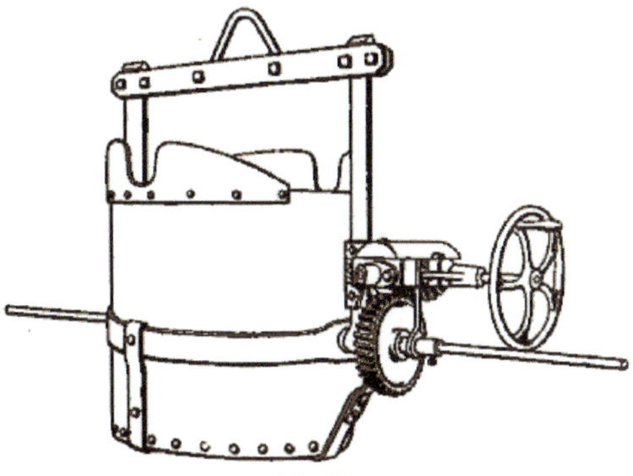

A gear driven ladle

In a foundry, a ladle is a vessel used to transport and pour out molten metals. Ladles range in size from small hand carried vessels that resemble a kitchen ladle and hold 20 kilograms (44 lb) to large steelmill ladles that hold up to 300 tonnes (330 tons). Many non-ferrous foundries also use ceramic crucibles for transporting and pouring molten metal and will also refer to these as ladles.

Types

The basic term is often pre-fixed to define the actual purpose of the ladle. The basic ladle design can therefore include many variations that improve the usage of the ladle for specific tasks. For example:

- Casting ladle: a ladle used to pour molten metal into moulds to produce the casting.
- Transfer ladle: a ladle used to transfer a large amount of molten metal from one process

to another. Typically a transfer ladle will be used to transfer molten metal from a primary melting furnace to either an holding furnace or an auto-pour unit.

- Treatment ladle: a ladle used for a process to take place within the ladle to change some aspect of the molten metal. A typical example being to convert cast iron to ductile iron by the addition of various elements into the ladle.

Unless the ladle is to be used with alloys that have very low temperature melting point, the ladle is also fitted with a refractory lining. It is the refractory lining that stops the steel vessel from suffering damage when the ladle is used to transport metals with high melting temperatures that, if the molten metal came in direct contact with the ladle shell, would rapidly melt through the shell. Refractory lining materials come in many forms and the right choice very much depends on each foundry's working practices. Traditionally ladles used to be lined using pre-cast firebricks however refractory concretes have tended to supersede these in many countries.

Foundry ladles are normally rated by their working capacity rather than by their physical size. Hand-held ladles are typically known as *handshank ladles* and are fitted with a long handle to keep the heat of the metal away from the person holding it. Their capacity is limited to what a man can safely handle. Larger ladles are usually referred to as geared crane ladles. Their capacity is usually determined by the ladle function. Small hand-held ladles might also be crucibles that are fitted with carrying devices. However, in most foundries, the foundry ladle refers to a steel vessel that has a lifting bail fitted so that the vessel can be carried by an overhead crane or monorail system and is also fitted with a mechanical means for rotating the vessel, usually in the form of a gearbox. The gearbox can either be manually operated or powered operation.

For the transportation of very large volumes of molten metal, such as in steel mills, the ladle can run on wheels, a purpose-built ladle transfer car or be slung from an overhead crane and will be tilted using a second overhead lifting device.

The most common shape for a ladle is a vertical cone, but other shapes are possible. Having a tapered cone as the shell adds strength and rigidity to the shell. Having the taper also helps when it comes time to remove the refractory lining. However straight sided shells are also fabricated as are other shapes.

The most common of these other shapes is known as a drum ladle and is shaped as a horizontal cylinder suspended between two bogies. Large versions, often having capacities in excess of 100 tonnes are used in steel mills are often referred to as torpedo ladles. Torpedo ladles are commonly used to transport liquid iron from a blast furnace to another part of the steel mill. Some versions are even adapted so that they can be carried on special bogies that can be transported by either road or rail.

Pour Designs

Ladles can be "lip pour" design, "teapot spout" design, "lip-axis design" or "bottom pour" design:

- For lip pour design the ladle is tilted and the molten metal pours out of the ladle like water from a pitcher.

- The teapot spout design, like a teapot, takes liquid from the base of the ladle and pours it out via a lip-pour spout. Any impurities in the molten metal will form on the top of the metal so by taking the metal from the base of the ladle, the impurities are not poured into the mould. The same idea is behind the bottom pour process.
- Lip-axis ladles have the pivot point of the vessel as close to the tip of the pouring spout as can be practicable. Therefore as the ladle is rotated the actual pouring point has very little movement. Lip-axis pouring is often used on molten metal pouring systems where there is a need to automate the process as much as possible and the operator controls the pouring operation at a remote distance.
- For bottom pour ladles, a stopper rod is inserted into a tapping hole in the bottom of the ladle. To pour metal the stopper is raised vertically to allow the metal to flow out the bottom of the ladle. To stop pouring the stopper rod is inserted back into the drain hole. Large ladles in the steelmaking industry may use slide gates below the taphole.

Ladles can be either open-topped or covered. Covered ladles have a (sometimes removable) dome-shaped lid to contain radiant heat; they lose heat slower than open-topped ladles. Small ladles do not commonly have covers, although a ceramic blanket may be used instead (where available).

Medium and large ladles which are suspended from a crane have a bail which holds the ladle on shafts, called trunnions. To tilt the ladle a gearbox is used and this is typically a worm gear. The gear mechanism may be hand operated with a large wheel or may be operated by an electric motor or pneumatic motor. Powered rotation allows the ladle operator to be moved to a safe distance and control the rotation of the ladle via a pendant or radio remote control. Powered rotation also allows the ladle to have a number of rotation speeds which may be beneficial to the overall casting process. Powered rotation obviously also reduces the effort required by the ladle operator and allows high volumes of molten metal to be transferred and poured for long periods without operator fatigue. Where the ladle is fitted with a manually operated gearbox, the type of gearbox most commonly used is the worm and wheel design because in most practical circumstances, and when correctly maintained it can be considered as "self-locking" and doesn't need an internal friction brake to regulate the tilting speed of the ladle. Other types of gear system can be also be used but they have to be fitted with an additional braking system that can hold the ladle if the operator takes his hand off the hand-wheel.

Lip-axis ladles may also use hydraulic rams to tilt the ladle. The largest ladles are un-geared and are typically poured using a special, two-winch crane, where the main winch carries the ladle while the second winch engages a lug at the bottom of the ladle. Raising the second winch then rotates the ladle on its trunnions.

Ladles are often designed for special purposes such as adding alloys to the molten metal. Ladles may also have porous plugs inserted into the base, so inert gases can be bubbled through the ladle to enhance alloying or metallic treatment practices.

What Modifications are Required?

Ladle is a cylindrical refractory lined vessel and aspect ratio of the bath varies between 0.8 and 0.9. This means that the bath is deep.

Bath agitation would be required to carry-out the functions effectively. At high temperature, bath can be agitated either by an inert gas or by induction. One has to determine the amount of stirring gas and location of the injection of gas in the ladle. The gas can either be injected through the nozzle or porous plugs. Location of the injection elements is an important issue. Injection elements could be located either axis-symmetric or asymmetric to the center of the ladle.

Enough free board height in the ladle must be available to accommodate the quantity of slag required for refining and to absorb inclusions.

Additional heating may be required to keep the molten steel to the teeming temperature. This can be achieved either by tapping steel at slightly higher temperature or to provide addition heating arrangement in the ladle itself.

In many situations it is required to inject the slag forming materials either for refining or for inclusion engineering. In this case suitable injection device must also be available.

Above all, the most important would be the selection of refractory to meet the refining requirements. Refractory materials for injection elements and their fixing must also be considered.

Basics of Gas Stirring

Argon is usually bubbled into the molten steel covered with slag either through the top lance or through a porous plug fitted at the bottom. A plume of gas rises upwards when gas is injected through the bottom. The stirring homogenizes bath composition and temperature. With centrally placed nozzle at the bottom, stirring action is small near the bottom of the ladle. An asymmetrically placed bubble plume gives velocities near the bottom which are greater than for symmetrically placed nozzle.

The liquid flow in the ladle occurs via bulk motion of the metal. Very roughly, characteristic velocity v_c for gas bubbling is

$$v_c = \left(\frac{QgH^2}{v}\right)^{1/3}$$

Where Q is gas flow rate $\left(\frac{m^3}{s}\right)$, g is $\left(\frac{m}{s^2}\right)$, H is bath height (m) and V is bath volume (m^3). For gas injection at 50 l/min in a ladle of bath diameter 3.5m and bath height 3.5m, equation calculates 0.14 m/s as the gas bubbling velocity.

At velocities greater than 0.3 m/s at the slag/metal interface slag droplets may be entrained by the metal flowing along the interface and into the melt.

For gas stirring recirculation rate of molten steel is of interest. The mass flux in (tons/s) of entrained steel passing through the top section of the bubble plume can be calculated by the following semi-empirical equation:

$$\dot{M} = 13.3(H + 0.8)\left[\ln\left(1 + \frac{H}{1.48}\right)\right]^{0.5} \times Q^{0.381}$$

Consider a ladle with bath height 3m and Q=650 l/min (1 atm and 273K). We can calculate $\dot{M} = 9.5\,tons/s$. Increasing the flow rate to 800 l/min increases $\dot{M}$ to 10.3 tons/s.

The calculations show that the recirculation time of a 200 ton melt is 21 s for 600 Nl/min and 20s for 800 Nl/min gas flow. At higher gas flow rates slag droplets may be entrained in molten steel which may increase the rates of refining reaction because of transitory phase contact .

For evaluating the efficiency of gas stirring, an alternative approach is to calculate energy input W in watt. The energy input can be calculated by

$$W = 6.18 \times 10^{-3} QT\left[\left(1 - \frac{273}{T}\right) + \ln\frac{P_1}{P_2}\right]$$

Here W is stirring energy in watt, Q is flow rate in Nl/min, t is bath temperature in K, P_1 is pressure at the bottom surface of the bath ($P_1 = P_2 + \rho_1 gH$; where $P_1$is density of molten steel, H is bath height and P_2 is atmospheric pressure). For gas injection rate 200 Nl/min into a bath of 2m height the stirring energy would be 390W/m³ when the bath diameter is 2.5 m.

Electromagnetic induction induces stirring energy about 100 W/m³. RH degassing gives about 800 W/m³ stirring energy at a circulating rate of 40 tons/minute.

Mixing time i.e. time to homogenize the bath indicates the conditions for stirring. Good mixing promotes the rate of slag/metal reaction as indicated by smaller mixing time. The following results are to be noted:

i) Scrap enhances the mixing time. Homogenization is possible within a definite time only up to a certain maximum scrap ratio.

ii) The mixing time decreases with increase in bottom gas rate

iii) Mixing time decreases with increasing the aspect ratio of bath.

iv) Mixing time(τ) in seconds can be correlated with the energy input in W/m³, volume of scrap (V_s):

$$\tau = [\exp(4.338 + 0.006\%V_s)] \times \left(W\frac{h_b}{d_b}\right)^{-0.00847\%V_s - 0.5564}$$

Where V_s in %.and $\frac{h_b}{d_b}$ expresses aspect ratio.

Gas Injection Rates

The following are some typical values of specific gas injection rates:

Stirring < 5Nl/ (min. t)

Lime steel desulphurization ≈ 14Nl/ (min. t)

Vacuum Arc Decarburization < 2Nl/ (min. t)

High gas flow rats though supply large amounts of stirring energy, large freeboards are needed. Large freeboards decrease the molten steel holding capacity of ladle. Lack of large freeboards and splashing often limits rate of gas injection.

Choice of Stirring Energy

Correct stirring is of utmost importance. It should be known when to stir what, and how vigorously. Vigorous stirring would be required for slag/metal reaction such as desulphurization. Inclusions removal would require weak stirring. Bath homogenization would also require weak stirring.

Vigorous mixing of metal and slag is achieved with gas stirring. Less disturbance at the slag/metal interface can be obtained by induction stirring. Slag carry-over from BOF/EAF must be avoided. Vigorous stirring in the neighborhood of slag/metal phase boundary activates interfacial mass transfer, leading to reduction of slag by deoxidizing elements, reversion of phosphorus, and oxygen and nitrogen pick up from the atmosphere.

Desulphurization of Steel

Synthetic slag practice can desulphurize steel up to 50% to 60% of original sulphur in steel. The following properties are desirable in synthetic slag:

i) Slag should have high sulphide capacity

ii) Basic slag is required

iii) Slag should be fluid to obtain faster reaction rates.

iv) Slag should not cause excessive refractory wear.

For efficient desulphurization , steel should be deoxidized and slag carry-over should be minimized. Fireclay ladles are not suitable if low sulphur steel is to be produced. Instead, dolomite or other basic refractory lined materials should be used. Argon bubbling is done.

Design of Synthetic Slag

The synthetic slag contains CaO, CaF_2, Al_2O_3 and with small amount of SiO_2. The principle component of synthetic slag is lime. Calcium fluoride increases the sulphide capacity of slag and helps fluidizing the slag. Often Al is present to deoxidize the molten steel since transfer of sulphur from molten steel to slag is followed by transfer of oxygen from slag to steel. Therefore deoxidation of steel is must for efficient desulphurization. Typically,slag contains $45-55\% CaO, 10-20\% CaF_2, 5-16\% Al$ and $0-5\% SiO_2$. This slag is pre fused in solid state.

Special synthetic slag can be designed for a specific purpose. For removal of oxide inclusions, a neutral slag with $CaO / SiO_2 = 1 or 1.2$ can be used, when no desulphurization is needed.

Issues Related to Synthetic Slag Practice

Synthetic slag practice appears to be simple and not much capital investment is needed. Certain issues are:

i) Desulphurization may vary from one heat to other if slag carry- over from BOF/EOF is not controlled. Oxygen content of steel should be same for consistent results.

ii) CaO is the main component. It is hygroscopic and leads to hydrogen pick up.

iii) Argon bubbling is done to stir the bath. Temperature drop could be of the order of $10^o C\, to\, 25^o C\; 150-250\, ton$ heat. The temperature drop is resulting from radiation heat loss from surface and heat transfer due to argon bubbling.

iv) The slag attacks the ladle refractory. Excessive amount of CaF_2 results in refractory wear. Higher tap temperature increases refractory wear.

Characterization of Synthetic Slag

An important parameter to characterize synthetic slag for its suitability to desulphurize molten steel is sulphide capacity of slag. On the basis of ionic theory, a modified sulphide capacity of slag is

$$C_s^1 = \frac{(W_s)\lfloor h_o \rfloor}{h_s}$$

Where (W_s) is % sulphur in slag and h_o and h_s are henrian activity of oxygen and sulphur in steel. In terms of property of slag sulphide capacity C_s is

$$C_s = (W_s) \times \sqrt{\frac{pO_2}{pS_2}}$$

Where (W_s) is weight percent of sulphur in slag with a gas having partial pressure of oxygen p_{O2} and p_{s2} . For a given slag, higher is the value of C_s, better is for desulphurization. C_s and C_s^1 are related with

$$\log C_s = \log C_s^1 + \frac{936}{T} - 1.375$$

At $T = 1823K$ and $1873K$, $\frac{C_s}{C_s^1} = 0.137$ and 0.133 respectively. Within the temperature range of desulphurization the ratio C_s / C_s^1 does not depend significantly on temperature. Alternatively C_s can be calculated by

$$\log C_s = 3.44(X_{CaO} + 0.1X_{MgO} - 0.8_{Al_2O_3} - X_{SiO_2}) - \frac{9894}{T} + 2.05$$

For a slag with $X_{CaO} = 0.6$ and $X_{SiO_2} = 0.4; C_s$ according to equation above is 2.87×10^{-3} at 1873.

Another important parameter is partition coefficient of S at equilibrium; it is defined as

$$K_s = \frac{(W_s)}{[W_s]} = \frac{C_s^i}{[h_o]}$$

h_o is activity of oxygen in steel and is determined by the amount of deoxidizer. Extent of desulphurization depends on extent of deoxidation. An aluminum killed steel can desulphurize steel much effectively that that when either $FeMn$ or $Fe\,Si$ is used to deoxidize steel.

Large value of K_s ensures efficient desulphurization, also large value of K_s requires low value of h_o and high value of C_s as well. By equation above

$$\log K_s = \log C_s^i - \log h[0]$$

$$\log K_s = \log C_s - \frac{936}{T} + 1.375 - \log h[0]$$

If aluminum is used to deoxidize steel, h_o in steel can be determined by

$$\log h_o = \frac{1}{3}\left[-\frac{64000}{T} + 20.57 - 20.57 - 2\log W_{al} + \log(a_{Al_2O_3})\right]$$

Combining equations above we get.

$$\log K_s = \log C_s + \frac{20397}{T} + \frac{2}{3}\log[W_{Al}] + \frac{1}{3}\log\left(a_{Al_2O_3}\right) - 5.482$$

Consider a slag with $X_{CaO} = 0.65$ and $X_{Al_2O_3} = 0.35$ which is used to desulphurize steel at $1873\,K$. Steel has $0.01\ wt\%Al$ dissolved. Activity of Al_2O_3 is 0.38 in slag. Let us calculate K_s by equation above.

$$\log C_s \text{ by equation } 4 = -1.96$$

$$\therefore \log K_s \text{ by equation } 9 = 97$$

Injection of Solid Powders

Injection techniques have the advantages of dispersing the reactants in the steel bath and at the same time provide a large reaction surface area. The type of powders used is governed by the purpose of injection. Table below shows the slag forming materials used for injection.

Purpose	Type of powder
Dephosphorization	$CaO+CaF_2+Fe_2O_3$+ mill scale soda
Desulphurization	CaO +Al $CaO + CaF_2$ +Al $Ca\ C_2$ Mg+(MgO,Al_2O_3, chloride slag) $CaC_2+ CaCO_3$, CaO

Alloying	Fe Si, $CaCN_2$, NiO, Mo O_2 Fe B, Fe Ti etc
Deoxidation and inclusion shape control	Al, Ca Si, Ca, Si, Mn, Al, Ca Si and Ba

Desulphurization Mechanism

Desulphurization can be carried out by injecting lime based powder, for example $90\% CaO+10\% CaF_2$ or $70\% CaO+20\% Al_2O_3+10\% CaF_2$. The injection rate varies between 2 -4 kg/ton of melt. When slag forming materials are injected into melt, they melt and the molten slag particles begin to rise and accumulate at the top surface of the melt. The desulphurization reaction occurs in two ways:

- During contact between rising molten slag particles and the melt. In this mechanism of desulphurization it is important that the powder becomes molten on injection. Residence time of the rising particles in the melt is also important, which means that the gas powder injection velocity must be suitably selected. Powder melts and the rising gas imparts mixing in the melt. This mechanism is known as transitory contact.
- Contact between top slag and the melt. As the molten slag particles rise they accumulate at the top surface of the melt and after a while top slag also takes part in the desulphurization. In this mechanism slag/metal interface area is important. Gas injection rate may be suitably selected to produce and entrain slag droplets into the melt for the faster rates of reaction. Once all the powder is injected, reaction between top slag and sulphur of melt governs the final sulphur content of steel. This mechanism is known as permanent contact.

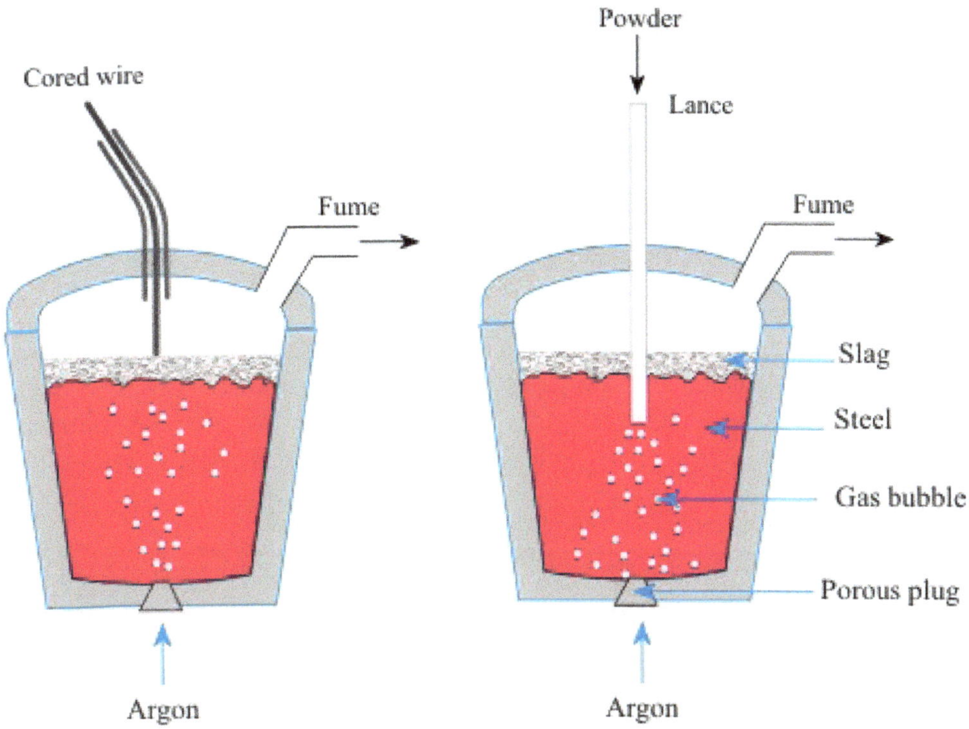

Ladle desulphurization carried out by powder injection technology (a) cored wire injection and (b) Pneumatic injection through a top lance

It must be noted that methods for injection of powder must also be developed. The slag forming materials are lighter than steel and deep injection would be required for the efficiency of the reaction. Powder can be injected either through cored wire or pneumatic transport. Figure shows arrangement of ladle desulphurization carried out either by injecting cored wire or by pneumatic injection through a top lance. In both argon is bubbled through a porous plug fitted at the bottom of the ladle. The desulphurization efficiency can be estimated by the mass balance.

In the transitory contact mode, the desulphurization can be determined by performing sulphur balance:

$$-1000\frac{d\lfloor m_s \rfloor}{dt} = \dot{W}_S L_S [m_S]$$

$\dot{W}_S$ is rate of injection of slag powder, m_S is sulphur in steel at time t. Integration of equation above and putting the limit

$t = 0, [m_S] = [m_S]_o$ and $t = t, [m_S] = [m_S]$ we get,

$$\frac{[m_S]}{[m_S]_o} = \exp\left[-\frac{\dot{W}_S L_S}{1000}\right] = \exp(-Z)$$

Similarly in the permanent contact zone sulphur balance

$$1000[m_S]_o + W(m_s)_o = 1000[m_s] + W_s(m_s)$$

Where W is amount of top slag, and m_s is initial S in slag and m_s is final sulphur in slag. Re- arrangement of equation above gives

$$R = 1 - \frac{\lfloor m_s \rfloor}{[m_S]_o} = \left\{\frac{L_S W_S}{1000 + L_S W_S}\right\} = \frac{1}{1+2}$$

Illustration

A slag with high value of L_S is desirable for efficient desulphurization. Consider two different slags whose partition coefficient is 80 and 500. Amount of slag injected is $4\ kg/ton$, $8\ kg/ton$, and 12 kg/ton. To illustrate the role of transitory and permanent contact on desulphurization , let use calculate $[m_S]/[m_S]_o$ for each case: The following table gives the calculated values

	L_o = 80 L_s = 500			
W_s kg/ton	$[m_s]/[m_s]_o$ trans.	$[m_s]/[m_s]_o$ perm.	$[m_s]/[m_s]_o$ trans.	$[m_s]/[m_s]_o$ perm.
4	0.73	0.14	0.14	0.33
8	0.53	0.018	0.018	0.20
12	0.38	0.0025	0.0025	0.14

We note the following:

- Both amount of slag and partition coefficient are important for desulphurization
- At lower amount of slag and lower partition coefficient there is not much difference in the ratio of $[m_S]/[m_S]_o$ for both permanent and transitory contact modes. For example for $L_s = 80$, when the powder injection rate is 4 kg/ton, the said ratio is 0.73 for transitory and 0.76 for permanent contact. However, when the powder injection rate is increased to 8 and 12 kg/ton, transitory contact mode becomes more efficient than permanent contact.
- At higher value of $L_s = 500$, transitory contact gives much lower value of sulphur in metal than permanent contact

Transitory contact mode requires designing powder injection systems which can inject powder at constant and uniform rate.

Alloying with Gas Injection

Alloying can be done during tapping by simply dropping the material on the surface, or with a carrier gas. The dissolution and homogenization of the alloying additions are enhanced by stirring and small particle size. Stirring intermixes top slag with the bath which should be minimized to avoid oxidation. The vigorous intermixing of top slag with the bath can be minimized by the lowering of a refractory lined cylinder into the liquid steel. This will minimize reaction of steel exposed to air.

Heating of Steel

Synthetic slag practice with argon stirring or injection of solid powder requires higher tap temperatures to compensate for the heat losses during refining. This increased tap temperature causes problems in BOF such as poor phosphorus removal and increased lining wear. In EAF increased power and electrode consumption and an increase in furnace time are the main issues to tap steel at higher temperature.

Arc heated ladle processes have been developed. Ladle arc furnace in addition to allowing for lower temperatures, also allows steelmakers to perform many metallurgical processes like:

i) Bath homogenization by argon stirring

ii) Inclusion removal and inclusion engineering

iii) Desulphurization by synthetic slag or by injection metallurgy

iv) Holding of ladles for long periods if and when need arises for example in sequence casting.

v) The ability to make addition of alloying elements.

There are three different types of ladle arc heated furnaces: Induction stirred, gas stirred by porous plugs and gas stirred using a tuyere.

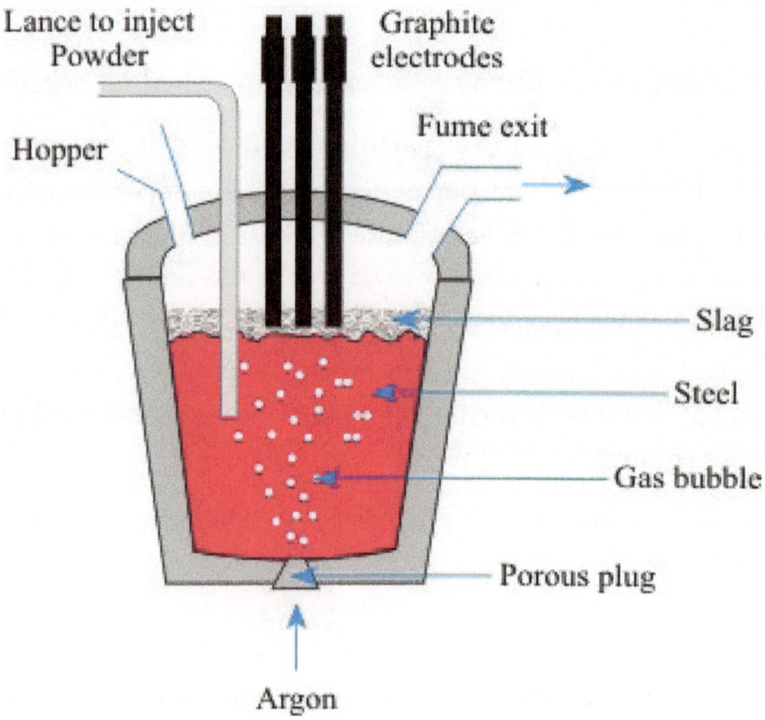

Ladle furnace showing arrangement of electrodes and porous plug

Figure shows ladle arc furnace where a cover equipped with three graphite electrodes are shown. The ladle bottom has a porous refractory plug for argon stirring. The ladle furnace is equipped with a hopper for additions of reactants. Fumes are taken out through the roof of the ladle.

The processes are under the following names:

ASEA- SKF: The process uses induction stirring. Induction stirring requires a stainless steel ladle or a stainless steel section of the ladle. Electrodes are used for heating.

Diado ladle furnace: The process is gas stirred. The furnace uses three graphite electrodes with submerged are heating.

FineKl- Mohr uses vacuum arc degassing (VAD) the system is under partial vacuum during heating. Argon bubbling is maintained through a porous plug.

Degasification

Degasification is the removal of dissolved gases from liquids, especially water or aqueous solutions. There are numerous possible methods for such removal of gases from liquids.

Gases are removed for various reasons. Chemists remove gases from solvents when the compounds they are working on are possibly air- or oxygen-sensitive, or when bubble formation at solid-liquid interfaces becomes a problem. The formation of gas bubbles when a liquid is frozen can also be undesirable, necessitating degassing. This process is called degasification.

Pressure Reduction

The solubility of gas obeys Henry's law, that is, the amount of a dissolved gas in a liquid is proportional to its partial pressure. Therefore, placing a solution under reduced pressure makes the dissolved gas less soluble. Sonication and stirring under reduced pressure can usually enhance the efficiency. This technique is often referred to as *Vacuum degasification*. Specialized vacuum chambers, called vacuum degassers, are used to degas materials through pressure reduction.

Thermal Regulation

Generally speaking, an aqueous solvent dissolves less gas at higher temperature, and vice versa for organic solvents (provided the solute and solvent do not react). Consequently, heating an aqueous solution can expel dissolved gas, whereas cooling an organic solution has the same effect. Ultrasonication and stirring during thermal regulation are also effective. This method needs no special apparatus and is easy to conduct. In some cases, however, the solvent and the solute decompose, react with each other, or evaporate at high temperature, and the rate of removal is less reproducible.

Membrane Degasification

Gas-liquid separation membranes allow gas but not liquid to pass through. Flowing a solution inside a gas-liquid separation membrane and vacuating outside makes the dissolved gas go out through the membrane. This method has the advantage of being able to prevent redissolution of the gas, so it is used to produce very pure solvents. New applications are in Ink Jet systems where gas in the ink forms bubbles and degrades print quality, a de gas unit is placed prior to the print head to remove gas and prevent the buildup of bubbles keeping good jetting and print quality.

The above three methods are used to remove *all* dissolved gases. Below are methods for more selective removal.

Substitution by Inert Gas

Bubbling a solution with an inert gas substitutes the dissolved harmful, reactive gases such as oxygen and carbon dioxide. Nitrogen, argon, helium, and other inert gases are commonly used. To complete the substitution, the solution should be stirred vigorously and bubbled for a long time. Because helium is not very soluble in most liquids, this technique is also used to reduce the risk of bubble formation in HPLC systems.

Addition of Reductant

If oxygen should be removed, the addition of reductants is sometimes effective. For example, especially in the field of electrochemistry, ammonium sulfite is frequently used as a reductant because it reacts with oxygen to form sulfate ions. Although this method can be applied only to oxygen and involves the risk of reduction of the solute, the dissolved oxygen is almost totally eliminated.

Freeze-pump-thaw Cycling

In this laboratory-scale technique, the fluid to be degassed is placed in a Schlenk flask and flash-frozen, usually with liquid nitrogen. Next a vacuum is applied, and the flask is sealed. A warm water bath is used to thaw the fluid, and upon thawing, bubbles of gas form and escape. The process is typically repeated three times. While this is a viable method for degassing a wide variety of organic solvents, some solvents are prone to significant expansion upon flash freezing. This expansion can cause practical problems by breaking the container. Some solvents known for such behavior are water and methanol.

Degassing Wine

Yeast uses sugar to produce alcohol and carbon dioxide. In wine making, carbon dioxide is an undesired by-product for most wines. If the wine is bottled quickly after fermentation, it is important to degas the wine first before bottling.

Wineries can often skip this step by aging their wines prior to bottling. Storing the wines in steel and wood barrels for months and sometimes years allows gases to be released from the juice and escape back into the air through air-locks.

Degassing

During steelmaking gases like oxygen, hydrogen and nitrogen dissolve in steel. The term degassing is employed to remove nitrogen and hydrogen from steel. Dissolved oxygen from steel melt cannot be removed as molecular oxygen.

Both nitrogen and hydrogen impair the mechanical properties of steel. The maximum solubility of nitrogen in liquid iron is 450ppm and less than 10ppm at room temperature. During solidification excess nitrogen is rejected which may form either blow holes or nitrides. Excess nitrogen causes embrittlement of heat affected zone of welded steels and impair cold formability. Hydrogen in steel impairs steel properties. Solubility of hydrogen in steel is low at ambient temperature. Excess hydrogen is rejected during solidification and results in pinhole formation and porosity in steel. Few ppm of hydrogen causes blistering and loss of tensile ductility. Thus removal of nitrogen and hydrogen from steel is necessary.

Principles

Consider removal of hydrogen and nitrogen from liquid steel

$$[H]=\frac{1}{2}\{H_2\}$$

$$[N]=\frac{1}{2}\{N_2\}$$

$$[Wt\%N]=\frac{p_{N_2}^{\frac{1}{2}}\times K_N}{f_N}$$

$$[Wt\%H] = \frac{p_{H_2}^{\frac{1}{2}} \times K_H}{f_H}$$

f_N and f_H are activity coefficient of nitrogen and hydrogen in steel, P_{N_2} and P_{H_2} are equilibrium partial pressure of nitrogen and hydrogen in steel.

$$\log f_N = e_N^i(\%i) + e_N^J(\%J) + ...$$

$$\log f_H = e_H^i(\%i) + e_H^J(\%H) + ...$$

Some values of interaction parameters are given below

Alloying elements i =	**C**	**Cr**	**Ti**	**P**	**Si**
e_H^i	0.045	+0.005	-0.22	0.011	0.027
e_N^i	0.13	- 0.045	-	0.051	0.047

$$\log k_N = -\frac{518}{T} + 2.937$$

$$\log k_H = -\frac{1905}{T} + 2.409$$

Let us calculate value of $[Wt\%H]$ when molten steel is degassed at 1850 K under vacuum. The pressure above the melt is reduced in one case to 1mm Hg and in other cases to 0.1 mm Hg and 10 mm Hg. Steel contains $C = 0.05\%, Cr\,6\%\,Ti\,0.6\%, Ni\,2\%$ rest iron. Assume $e_H^{Ni} = 0$.

$$\log f_H = 0.05 \times 0.045 + 0.005 \times 6 + 0.22 \times 0.6$$

$$f_H = 0.795$$

$$k_H = 23.94$$

Pressure (mmHg)	[ppm H]
0.1	0.218
1.0	1.092
10	3.450

We note that with the decreasing pressure above the melt dissolved hydrogen in steel decreases. At 10 mm Hg pressure above the melt, dissolved hydrogen is 3.450 ppm which decreases to 0.218 ppm at 0.1mm Hg pressure, which is 6% of hydrogen at 10mm Hg pressure. Higher degree of vacuum is beneficial with reference to degassing. But higher degree of vacuum requires proper selection of ladle refractory material to avoid decomposition of the refractory and vacuum equipment. Also side reactions may occur at higher degree of vacuum.

Side Reactions

During vacuum degassing the following reaction may occur. Nitride and oxide inclusions can decompose according to

$$Al\,N = [Al] + [N]$$

Application of vacuum decreases nitrogen which favors decomposition of $Al\,N$. Oxide inclusions can react with C

$$SiO_2 + C = \{Si\} + 2\{CO\}$$

Lowering of CO pressure favours the forward reaction. Reaction between lining and carbon of liquid steel or decomposition of lining may occur:

$$SiO_2 + C = \{Si\} + 2\{CO\}$$

$$MgO = Mg(g) + [O]$$

$$CaO = Ca(g) + [O]$$

$$MgO + C = Mg(g) + \{CO\}$$

Note that SiO is a highly reactive gas at the steelmaking temperature. All reactions generate one or more gaseous species, hence lowering of pressure promotes the occurrence of side reactions. Mg and Ca are stable gases at steelmaking temperatures.

Volatilization of elements of high vapor pressure may occur. Mn and Fe have high vapor pressures and their losses occur during vacuum treatment. Loss of Al and Si is negligible.

General considerations

1. The desorption of gases is a gas/ metal interfacial reaction. The atomic nitrogen from the molten steel has to diffuse at the gas/metal interface, where it is converted to molecular nitrogen which can then be desorbed. The effectiveness of vacuum treatment increases with increase in surface area of liquid exposed to vacuum. The increased surface area of molten steel exposed to vacuum e.g. in the form of a thin stream or gas induced stirring will accelerate the degassing process.

2. Temperature of molten steel drops during vacuum treatment. More is the surface area of stream exposed to vacuum higher will be the temperature drop. To compensate for the loss of heat, the following alternative may be considered.

Tapping of steel at a higher temperature. This requires increased heat load in BOF/EAF

Additional heating during vacuum treatment.

3. The degassing time must be kept to minimum.

4. The degree of degassing increases with the degree of vacuum. Vacuum of the order of 1mm

or even less than 1mm Hg (1mmHg=1torr) is employed in the practice. Vacuum pumping capacity should be adequate.

Fluid Flow in Degassing

Degassing can be carried out either by placing ladle containing molten steel under vacuum or by recirculation of molten steel in vacuum.

In ladle degassing, the effectiveness of degassing decreases from top to bottom of the molten steel bath. Bottom layers of steel are very much less affected by vacuum since these layers are under the influence of ferrostatic pressure due to column of liquid steel. Hence bath agitation would help exposing the entire content of molten steel to the vacuum. Argon stirring is commonly employed during degassing. Argon bubbling during degassing of molten steel leads to massive volumetric expansion of bubbles due to temperature.(volume of gas becomes 6.3 times at 1873K). Moreover rising gas bubbles absorb dissolved gases. Radial expansion of gas bubble in vacuum processing impart to a radial motion to the surrounding fluid.

In recirculation degassing argon is also bubbled through porous plugs located at the bottom of the ladle. The speed of degassing increases with the increased rate of circulation (R) of liquid steel through the vacuum chamber. Typically R ranges in between 10t/min to 100t/min. Circulation velocity increases with an increasing argon gas flow rate.

The circulation rate (R) can be determined by

$$R = 7.42 \times 10^3 Q^{1/3} d^{1/3} \left\{ 1n\left(\frac{p1}{p2} \right) \right\}$$

Here R= circulation rate Kg/s

Q= argon injection rate $Nm^3 / \min$.

p_1 = Pressure at the base of downleg

p_2 = Pressure in vacuum chamber

D= internal diameter of leg (m)

Material balance in degassing

Consider removal of hydrogen in recirculation degassing i.e. during circulation of molten steel in vacuum.

Hydrogen Balance

Rate of hydrogen removal from steel $(\dot{M}_1)$= rate at which hydrogen is transferred in vacuum $(\dot{M}_2)$ If weight of steel is m in tones, t is time in minutes and ppmH is hyd

$$(\dot{M}_1) = -m \times 10^6 \frac{d[PPmH] \times 10^{-6}}{dt} = -m \times \frac{d[PPmH]}{dt}$$

$$\left(\dot{M}_2\right) = R\times 10^6 \{[PPmH]-[PPmH]^*\}\times 10^{-6}$$

R is circulation rate in tone /min, and $[PPmH]^*$ is hydrogen concentration at equilibrium. By equations above

$$\frac{R}{m}dt = -\frac{d[PPmH]}{[PPmH]-[PPmH]^*}$$

Integrating equation above and using boundary conditions

$$\lfloor PPmH \rfloor = \lfloor PPmH \rfloor_1 \ at\ t = 0$$

$$[PPmH] = [PPmH]_2 \quad at\ t = t$$

$$R = \frac{m}{t}\ln\frac{\lfloor PPmH \rfloor_1 - \lfloor PPmH \rfloor^*}{[PPmH]_2 - [PPmH]^*}$$

Degassing Processes

There are 3 methods of degassing which are in practice:

1. Ladle degassing
2. Stream degassing
3. Circulation degassing

All these processes are carried out in ladles.

Ladle Degassing

Ladle containing molten steel is placed in a chamber which is then evacuated. After a determined time ladle is removed from the chamber and is teemed for casting. Figure shows ladle degassing unit.

Ladle is provided with a porous plug at its bottom to purge argon gas as shown in the figure. In a vacuum chamber the ladle is placed. The vacuum chamber is equipped with a hopper so as to make additions of elements as and when it is needed. For effective degassing of fully killed steel, it is necessary to purge argon through the bottom of the ladle. Stirring the bath enhances rate of gas removal. Vigorous removal of gases causes metal splashing too. Therefore ladle is not filled completely and about 25% of its height is kept as freeboard to accommodate the splashed metal droplets. Pressure is maintained in between 1mmHg to 10mm Hg for effective degassing. During degassing additions are made for deoxidation and alloying. In certain cases ladle is heated to compensate for the loss of heat during degassing. For the effectiveness of degassing , it is necessary that carry-over slag either from BOF or EAF should be as low as possible. Carry-over slag contains FeO and since oxygen content of steel is in equilibrium with FeO content of slag, oxygen content of steel increases.

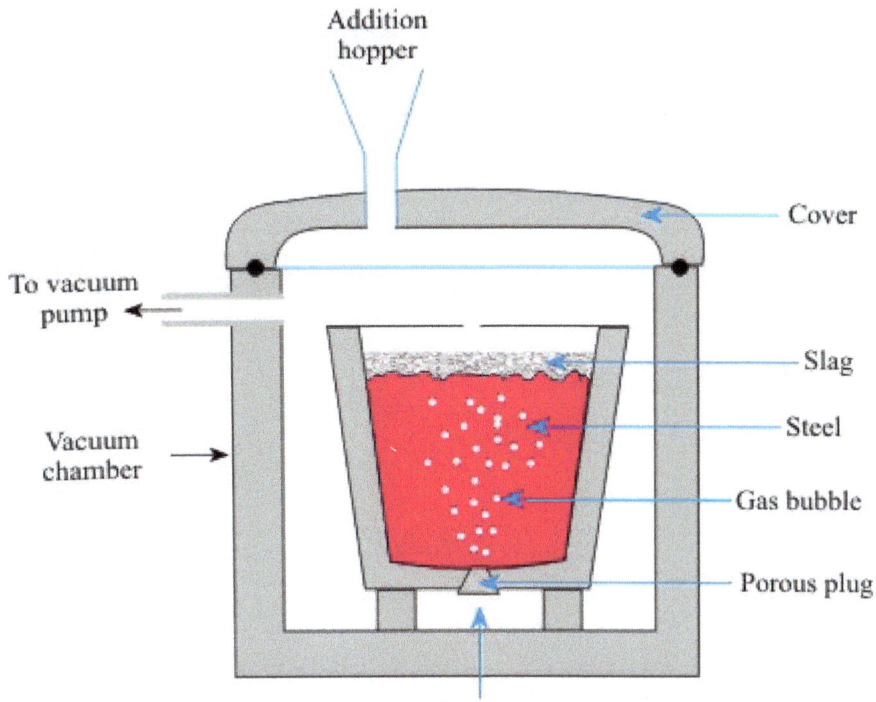

Arrangement of ladle with porous plug and hopper for degassing

Stirring gas is introduced either from top through the roof by a submerged refractory tube or through the porous plug fitted at the bottom of the ladle.

Electromagnetic stirring is employed for degassing. For this purpose ladle has to be made of non magnetic austenitic stainless steel or stainless window could be provided.

For certain grades of alloy steels, both induction stirring and arc heating are employed for degassing.

The final content of gas in steel depends on degree of vacuum and time of treatment. Hydrogen is generally reduced to below 2ppm from 4 to 6ppm, nitrogen content of steel is also reduced. The pick-up of nitrogen from the atmospheric air may occur during open pouring of steel, which must be controlled.

Ladles are generally lined with high alumina bricks at upper part of the ladle while the lower portion is lined with fireclay.

Stream Degassing

In stream degassing technology, molten steel is teemed into another vessel which is under vacuum. Sudden exposure of molten stream in vacuum leads to very rapid degassing due to increased surface area created by breakup of stream into droplets. The major amount of degassing occurs during the fall of molten stream. Height of the pouring stream is an important design parameter.

Stream degassing technology has following variants in the practice

i Ladle to Mould Degassing

Preheated mold with hot top is placed in vacuum chamber. Above the chamber a tundish is placed. Steel tapped in the ladle at superheat equivalent to $30^{\circ}C$ is placed above the tundish. Steel is bottom poured in the tundish. One ingot could weigh around as high as 400tons and several heats from different furnaces are used for casting. Figure shows arrangement of vessels.

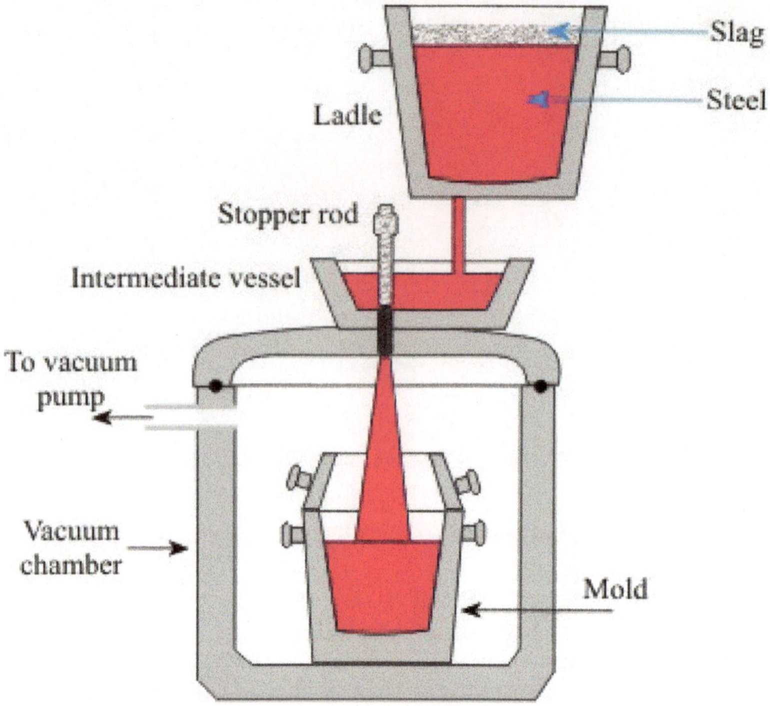

Arrangement of ladle, tundish and mold to degass molten steel

ii Ladle to Ladle Degassing

In ladle to ladle degassing, a ladle with the stopper rod is placed in a vacuum chamber. Ladle containing molten steel from BOF or EAF is placed on top of the vacuum chamber and the gap is vacuum sealed. Alloy additions are made under vacuum. Stream is allowed to fall in the ladle where molten steel is degassed. Alloy additions are made under vacuum.

In some plants degassing is done during tapping. In this arrangement molten steel from EAF is tapped into tundish or pony ladle. From the pony ladle molten stream is allowed to fall into a ladle which is evacuated. Ladle is closed from top with a special cover which contains exhaust opening. Steel with 25% to $30^{\circ}C$ superheat is tapped into ladle.

Recirculation Degassing

In the recirculation degassing technology, molten steel is allowed to circulate in the vacuum chamber continuously by special arrangement.

In RH degassing technology (developed by Rheinstahl Heinrich – Shutte at Hattingen, Germany),a cylindrical refractory lined shell with two legs (also called snorkel) is designed such that steel is raised in one leg and falls back into the ladle after degassing through the second leg. Top side of the cylindrical shell is provided with exhaust, alloy additions, observation and control window.

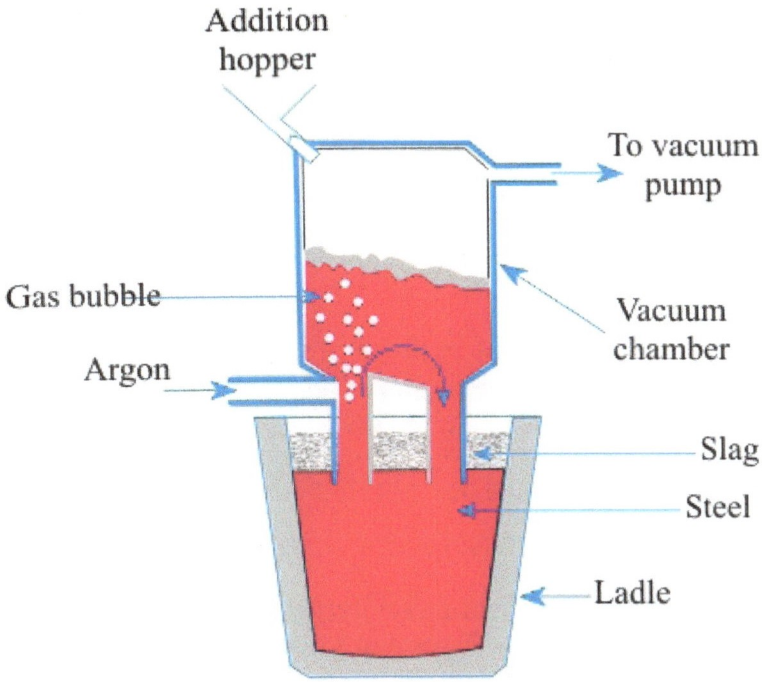

Arrangement of cylindrical vessel and ladle in RH degassing technology

Cylindrical Shell is lined with fire bricks in the upper portion, and alumina bricks in the lower portion in order to sustain high temperature. The legs are lined with alumina refractory. A lifter gas argon is injected at the inlet snorkel in order to increase the molten steel velocity entering into inlet snorkel. Figure shows a schematic sketch of a RH degassing unit.

The operation of RH degasser is as follows:

i) Cylindrical chamber is heated to the desired temperature (varies in between $900^{o}C$ to $1500^{o}C$ in different plants).

ii) The chamber is lowered into molten steel up to a desired level.

iii) The chamber is evacuated so that molten steel begins to rise in the chamber. Lifter gas is introduced. This gas expands and creates a buoyant force to increase the speed of molten steel rising into the inlet snorkel.

iv) Molten steel in the chamber is degassed and flows back through the other snorkel into the ladle. This degassed steel is slightly cooler than steel in the ladle. (Buoyancy force created by density difference density of cooler liquid steel is > hot steel) stirs the bath.

v) Rate of circulation of molten steel in cylindrical chamber controls the degassing. Circulation rate depends upon amount of lifter gas and the degree of vacuum. A 110 T steel can be degassed in 20 spminutes by circulating molten steel at 12 tons/min, amount of argon is around $0.075 to 0.075 m^3 / ton$.

vi) Alloy additions can be made at the end of degassing depending on the superheat.

vii) Process has several advantages like

- Heat losses are relatively low.
- Alloy additions can be adjusted more closely
- Small vacuum pumping capacity is adequate since smaller volume is to be evacuated as compared with ladle to ladle or stream degassing.

DH Gassing

In DH degassing, a small amount 10-15% of the total mass of steel is degassed at a time. The process is repeated until required level of degassing is achieved. The arrangement of a vessel and the ladle is somewhat similar to figure except the following:

- In DH unit, the cylindrical vessel has one snorkel .
- Cylindrical vessel has heating facility.

The DH chamber is equipped with heating facility, alloying addition arrangement and exhaust systems. Bottom of the cylindrical vessel is provided with a snorkel which can be dipped into molten steel. The upper portion of the DH vessel is lined with the fireclay and the lower portion with the alumina bricks, snorkel is lined with high quality alumina brick. The length of the snorkel is sufficiently large to realize the effect of atmospheric pressure on rise of steel in the snorkel. The following steps may be noted for operation:

- DH vessel is preheated and lowered in the ladle so that snorkel tip dips below the molten steel surface
- The evacuated chamber is moved up and down so that steel enters the chamber
- The chamber is moved for 50-60 times with a cycle time of 20 seconds.
- Adequate degassing is possible in 20 -30 cycles.
- A layer of slag is kept in the ladle to minimize heat losses.
- The DH degassing unit can operate with lower superheats compared with RH since DH unit has heating facility

Clean Steel

What is Clean Steel?

Clean steel refers to steel which is free from inclusions. Inclusions are non metallic particles embedded in the steel matrix. Practically it is not possible to produce steel without any inclusion. Therefore we can talk about cleaner steel. Which steel is clean would depend on the applications.

In this connection it is important to know that there is a limiting size below which inclusion does not affect mechanical property. Table lists some applications which can tolerate some minimum inclusions size:

Steel product	Allowed impurity in ppm	Allowed size μm
Automotive and deep drawing sheet	C<30, N<30	100
Line pipes	S<30, N<50, TO<30	100
Bearings	TO<10	15
Tire cord	H<2N<40, TO<15	10
Heavy plate steels	H<2 N+30 to 40, TO<20	13
Wires	N<40, TO<15	20
Drawn and ironed cans	C<30, N<40, TO<20	20

Types of Inclusions

Inclusions are chemical compounds of metals like (Fe, Mn, Al, Si, Ca etc) with non metals (O, S, N, C, H). Different types are:

- Oxides: $FeO, Al_2O_3, SiO_2, MnO, Al_2O_3.SiO_2, FeO.Al_2O_3, MgO.Al_2O_3, MnO.SiO_2$
- Sulphides: $FeS, CaS, MnS, MgS, Ce_2S_3$,Nitirides: TiN, AlN, VN, BN etc.
- Oxysulphides: $MnS.MnO, Al_2O_3.CaS$, etc.
- Carbonitrides: Titanium/ vanadium/Niobium carbonitirides, etc.
- Phosphides: Fe_3P, Fe_2P, Mn_5P_2

By mineralogical content, oxygen inclusions are classified:

- Free oxides - $FeO, MnO, Cr_2O_3, Sio_2(quartz) Al_2O_3$ (corundum) and other;
- Spinels – Ferrites, chromites and aluminates.
- Silicates- SiO_2 with a mixture of iron, manganese, chromium, aluminum and tungsten oxides and also crystalline silicates.

By stability, non –metallic inclusions are rather stable or unstable. Unstable inclusions are iron and manganese sulfides and also some free oxides.

Morphology

Globular shape is desirable: Certain inclusions like MnS, oxysulphides, iron aluminates and silicates are globular.

Platelet shape: undesirable. Al deoxidized steels contain MnS in the form of thin films located along the grain boundaries.

Polyhedral inclusions are not very harmful.

Size of inclusions

There are micro inclusions ($size 1-100\mu m$) and macro-inclusions (size greater than $100\mu m$). Macro inclusions are harmful. Micro inclusions are beneficial as they restrict grain growth, increase yield strength and hardness. Micro- inclusions act as nuclei for precipitation of carbides and nitrides. Macro- inclusions must be removed. Micro inclusions can be used to enhance strengthening by dispersing them uniformly in the matrix.

Properties of Inclusions

i) Thermal expansion

An inclusion is a mismatch with the steel matrix. There are inclusions like MnS,CaS etc. which have thermal expansion greater than steel matrix. On heating steel with these types of inclusions voids or parting of the matrix can occur. The void can act as cracks.

On the other hand Al_2O_3, SiO_2 and $CaO.Al_2O_3$, etc inclusions have thermal expansion smaller than steel matrix. On heating steels with these type of inclusions internal stresses of thermal origin can develop.

Density and Melting Point

ii) Density and melting point

Composition of inclusions	Melting point(°C)	Density at 20°C(g/cm³)
Ferrous oxides (FeO)	1369	5.8
Manganous oxides (MnO)	1785	5.5
Silica	1710	2.2-2.6
Alumina (Al_2O_3)	2050	4.0
Chrome oxide Cr_2	2280	5.0
Titanium oxide, TiO_2	1825	4.2
Zirconium oxide, ZrO_2	2700	5.75
Iron silicate, $(FeO)_2SiO_2$	1205	4.35
Iron sulphide, FeS	988	4.6
Manganese sulphide, MnS	1620	4.04
Magnesia, MgO	2800	3.58

Inclusions like MgO, Al_2O_3, TiO_2 are solid at steelmaking temperature.

iii) Plastic deformability

The plastic deformability of an inclusion will govern any change in its shape under the action of external forces and will determine the amplitude of stress concentration. Brittle inclusions are dangerous as they may crack and cause fracture of the component under the application of external force. The majority of inclusions belong to pseudo-ternary system: $CaO-SiO_2-Al_2O_3, MgO-SiO_2,-Al_2O_3$ $CaO-SiO_2-Cr_2O_3$ etc. Sulphide inclusions are mainly MnS. Other elements like Ti, Zr, rare earths, Nb, V etc. usually appear as solid solutions in existing inclusion phases. The following classification of inclusions according to kieslling is useful to the metallographers to determine type of inclusions: According to Kieslling:

- Calcium aluminates and Al_2O_3 inclusions in steel are undeformable at temperatures of interest in steelmaking.
- Spinel type double oxides AOB_2O_3 (where A is Ca, Fe(l), Mg and Mn,and B is Al,(Cr etc) are deformable at temperatures greater than $1200^o C$.
- Silicates are deformable at higher temperature range. The extent of deformation depends on their chemical compositions. Silicates are not deformable at room temperature.
- FeO, MnO and $(Fe, Mn)O$ are plastic at room temperature but gradually lose plasticity above $400^o C$.
- MnS which is highly deformable at $1000^o C$ temperature but becomes slightly less deformable above $1000^o C$.
- Pure silica is not deformable up to $1300^o C$.

Inclusion Assessment

Inclusion counts are performed to assess their shape, quantity and distribution to assess about the cleanliness of steel.

The routine plant procedure employs the microscopic method. From the shape of the inclusion and knowledge of the steelmaking process in a plant, it is inferred to whether it is silica/ silicate, aluminate or sulphide inclusion.

Electron probe micro analyzer enables to determine the chemical composition of individual inclusions.

The energy dispersive x-ray analysis (EDX) attachment for SEM allows quantitative chemical analysis of inclusion as well as quantitative mapping of distribution of various elements in and around the inclusions.

Quantimet has an optical microscope fitted with video screen and associated microprocessor-based instrumentation. It can scan the specimen very quickly and provide a variety of information such as inclusion size, distribution, number, volume fraction, etc.

Total oxide inclusion content of steel can be determined from the analysis of oxygen by sampling and the use of vacuum /inert gas fusion apparatus.

Radioactive tracers can identify the origin of inclusion distributions, etc.

Why Inclusion Control is Necessary?

Impact properties are adversely affected with an increase in volume fraction of inclusion as well as inclusion length; spherical inclusions are better. Brittle inclusions or inclusions that have low bond strength with the matrix break up during early straining and create voids at the inclusion/ matrix interface.

Hot fatigue strength of high strength steel is reduced by surface and subsurface inclusions those have lower thermal expansion coefficient than steel

The hot workability of steel is affected by the low deformability of inclusions.

Anisotropy of a property is caused by orientation of elongated inclusions along the direction of working. Macro inclusions of sulphides are desirable for better steel machining properties.

In view of the above, control of inclusion in steel is necessary. It must be mentioned and as pointed out earlier macro- inclusions are harmful and they must be controlled.

Sources of Inclusion Formation

Inclusions can form either (a) during transfer of molten steel from one reactor to other or (b) during solidification of steel or during solid state processing by any of the following mechanisms:

i) Reaction between rejected solute elements during solidification, for example, reaction between sulphur and manganese, and between oxygen and aluminium etc.

ii) Mechanical and chemical erosion of refractory and other materials.

iii) Oxygen pickup by teeming stream and consequent oxide formation.

iv) Chemical reactions.

Inclusions produced by mechanism (i) are called endogenous, whereas mechanisms (ii) and (iii) produce exogenous inclusions. Inclusions can form during

a) Tapping of molten stream from BOF/EAF to ladle. Erosion of launder refractory is the possible source. Pick up of oxygen from atmosphere and formation of FeO.

b) Treatment of steel in ladle. Here molten steel is in contact with the refractory. Also during deoxidation and synthetic slag treatment oxide/sulphide inclusions may form.

c) Teeming of molten stream. Molten steel is in contact with stopper and nozzle refractory and elements like Ti, Mg etc., which can form oxides. Air entrainment into molten steel stream brings oxygen and FeO formation is initiated.

d) Solidification in mould due to precipitation of excess solute elements.

e) Final finishing operations like heat treatment and deformation processing. Here steel is heated to high temperature which may cause surface oxidation, surface sulphurization, inner oxidation,etc.

f) Fusion welding. Oxidation of weld pool, electrode coatings are the possible source of inclusions in fusion welding processes.

Control of Inclusions

Inclusions can be controlled either at (a) during liquid state processing stage or (b) during solid state processing.

Liquid State Processing

During liquid state processing inclusion control can be exercised at tapping and teeming of steel.

i) During tapping of molten stream from BOF/EAF carry- over of slag must be minimized if not prevented. BOF slags are highly oxidizing in nature and contain oxides like Fe, MnO, SiO_2, CaO, MgO etc. These oxides react with Al during ladle treatment and lead to inclusion formation. Consider the reaction

$$3\,FeO(1) + 2\,Al = Al_2O_3 + 3\,Fe$$

$$3\,MnO(1) + 2\,Al = Al_2O_3 + 3\,Mn$$

Carry over of 1 kg FeO in slag decreases Al by 0.25 kg. which in turn forms 0.47kg Al_2O_3. Assuming spherical shape inclusion of 1mm dia (in practice different diameter of inclusions can form) and density of $Al_2O_3 = 4000\frac{kg}{m^3}$, number of inclusions are 224/kg carry-over FeO of slag. Thus carry-over of slag must be minimized by adopting slag free tapping technologies.

Another way for inclusion formations is reoxidation of tapping stream. Tapping stream exposes very large surface area in the atmosphere and hence oxygen pick up leads to oxide inclusion formation.

ii) Molten steel stream after treatment in the ladle is teemed into tundish and then from tundish to mould in the continuous casting.

Teeming of steel form ladle to tundish requires shrouding of molten steel stream in order to avoid reoxidation. Here macro-inclusions rich in FeO and MnO can form. Moreover, dissolved oxygen increases and forms inclusion during solidification. Use of shrouded and submerged nozzles will help control inclusion formation. Different technologies for shrouding molten stream are: refractory tube shroud, circular ring shroud, etc.

Teeming of molten steel from tundish to continuous casting mould needs extra precaution in terms of protection of steel against atmosphere. Since oxygen pick up occurs just before the solidification, resulting inclusions do not get sufficient time for separation owing to faster solidification in mold Teeming stream by inert gas or through the use of nozzles submerged in molten stream are the effective means to avoid inclusion formation.

iii) Selection of tundish flux

Tundish flux should be selected such that it can easily absorb inclusions floating in the tundish. At the same time flux should also cover molten steel to prevent oxidation.

iv) Tundish operation

Now a days sequence casting is commonly adopted in continuous casting. During ladle change-over period tundish feeds the molten steel to different molds of the continuous casting machine. In this situation care must be exercised to ovoid slag entrainment into mold due to vortex formation. Tundish should not be emptied completely. Also tundish lining material should be inert with aluminium.

v) Inclusion can also form in the mold during solidification. As the steel solidifies the excess solute elements like oxygen, sulphur, manganese etc are rejected and lead to inclusions formation.

Segregation during solidification must be avoided. Here stirring of molten steel is effective to minimize the segregation and inclusion formation.

Solid State Processing

In the solid state processing steel is heated to a temperature ranging in between 800-1200°C to perform heat treatment and hot working. Here steel must be heated under inert atmosphere to avoid oxidation. During fusion welding, liquid pool is in contact with air and steel is prone to oxidation. Inert shielding may avoid the inclusion formation.

Inclusion Engineering

Inclusions engineering technology is based on the design of inclusions so as to alleviate their harmful effects on the product properties.

Inclusion engineering does not refer to removal of inclusions but it refers to modify them either in terms of chemical composition or shape so that harmful effects of the inclusions can be converted to improve the steel properties.

Inclusion engineering also involves distribution of inclusion uniformly in the matrix. So that composite properties can be generated in the product.

In some cases, deliberate attempts are made to form very fine inclusions (e.g. nitrides, and carbonitrides inclusions in hardening steel). Such inclusion can form by reaction between Ti, W,Al with O, N, S or C.

Macro-size inclusions must be removed. In all other cases, depending on applications, inclusion can be modified to minimize their harmful effects.

Requirements for Inclusion Engineering

One of the main requirements is to find an element which should be added to modify the inclusion. Element should meet the following:

- It should have high chemical affinity for the inclusion.
- It should be able to modify the composition so that it becomes liquid.
- It should also be able to modify the shape i.e. sharp edges ands corner of inclusions to spherical.

What Types Of Inclusions Need To Be Modified?

Mainly Al_2O_3 and MnS inclusions are modified. Al_2O_3 inclusions are brittle and breaks on deformation. Similarly Al_2O_3 inclusion in steel on deformation becomes stringer type.

Inclusion Engineering

In aluminum killed steels, alumina inclusions are formed. Alumina inclusions are solid at steelmaking temperature. During continuous casting, alumina inclusions clog the nozzle and consequently steel flow rate to the mold is affected. Calcium is used widely to modify inclusions in continuous casting of steel. The solubility of calcium in steel is 320 ppm. The density of calcium is 1.55g/cm³ and is in vapour form at steelmaking temperature $(1600^o C)$. Therefore suitable injection methods for calcium have to be developed.

One method is to inject calcium deep into the molten bath such that the ferrostatic pressure overcomes vapor pressure of calcium. In another method wire containing calcium components are injected at speeds of 80-300 m/min. Typically Ca-Si powder in which calcium content is 20-30% is used in the wire. Wire containing Ca-Si powder is injected deep into bath to avoid a) oxidation of Ca and b) to keep calcium in the liquid form.

Addition of calcium in steel containing oxygen and sulphur can form two phases: oxide and sulphide. Oxide phase consists of the compounds in $CaO - Al_2O_3$ system. The different oxide compounds have different melting temperature as shown in the table.

Compound	Melting point (K)	Ca/Al
3 CaO. Al_2O_3	1800	2.22
12 CaO.7 Al_2O_3	1673	1.27
CaO. Al_2O_3	1868	0.74
CaO. 2 Al_2O_3	2000	0.37

It may be noted that $12CaO.7\,Al_2O_3$ has lowest melting point and it will remain in the liquid at steelmaking temperature. The table also shows Ca/Al ratio in various $CaO - Al_2O_3$ compounds. The Ca/Al ratio is 1.27 in $12CaO.7\,Al_2O_3$ compounds. This suggests that calcium/aluminum ratio has to be adjusted at 1.27 (theoretically) in order to obtain liquid product.

The sulphide phase consists of CaS and MnS. The melting point of CaS is $2000^o C$ and that of MnS $1610^o C\,CaO.2\,Al_2O_3$ melt has some solubility for CaS.

Oxide Shape Control

It consists of modifying Al_2O_3 inclusions. Al_2O_3 is solid at steelmaking temperature and is brittle in nature. On rolling, alumina inclusions break up which is a serious surface defect. The alumina may accumulate in the continuous casting nozzle causing clogging of the nozzle.

For a normal aluminum killed steel calcium will first modify the oxide inclusions. Thermodynamically it is easier to form $CaO.2\,Al_2O_3$ type of inclusions which will be converted to $CaO.2\,Al_2O_3$ and finally liquid calcium aluminate rich is CaO. The calcium will then desulphurize steel to very low levels.

Thus it is possible that calcium will react with sulphur to form CaS before there is oxide shape modification. The sulphur level where CaS is more stable will depend on Al content. It must also be noted that $12CaO.7\,Al_2O_3$ is liquid product at steelmaking temperature. Thus it is important to control the concentration of Al, S and Ca in molten steel for engineering of oxide inclusions.

Sulphide Shape Control

MnS has a low melting point and elongate in the direction of rolling and becomes stringer type inclusions. These inclusions greatly reduce the transverse mechanical properties of steel. MnS also causes steel to be more susceptible to hydrogen induced cracking. Stringer type inclusions should be modified if not eliminated.

Formation of MnS requires thermodynamically that the solubility product

$$[\%Mn]\times[\%S]>2$$

at the solidification temperature. During solidification, Mn and S are rejected from the solid steel causing an increase in their concentration in the remaining liquid. The Mn and S enriches in the interdendritic region and MnS forms during the end of solidification.

For sulphide shape control, it is necessary to desulphurize steel to very low value say 0.006%. Addition of calcium to aluminum killed steel will first form calcium aluminate inclusion. If calcium aluminate inclusion has excess CaO content, it has high sulphide capacity as well. This will result in precipitation of CaS which forms a duplex inclusion in which a $CaS-MnS$ ring surrounds calcium aluminate core. This type of inclusion is spherical and does not elongate.

Effect of Rare Earth Elements

When rare earths (elements) like cerium, lanthanides are added to steel they can react to form an oxide, sulphide or oxysulphides:

$$2Ce+3O=Ce_2O_3$$

$$Ce+S=CeS$$

$$2Ce+2O+S=Ce_2O_2S$$

Correct addition of rare earth elements like cerium forms oxysulphide inclusion.

It must be noted that rare earths have high atomic weights, and their oxides, sulphides and oxy-sulphides have high density 5 to 6 g/cm^3 as compared with other oxides. These inclusions do not float easily. Rare earths are strong deoxidizers and desulphrizers as well.

As an example of inclusion engineering, oxide inclusions have been designed to make them useful to increase the machine tool life. Subramanian et al has designed tool steel to be self lubricating by modifying the rheology of oxide inclusions. Glassy inclusions are designed to soften at the tool/chip interface temperature so that a viscous layer of glass can lubricate the tool-chip interface.

5

Solidification and Casting of Steel

After the removal of superheat and latent heat, liquid changes into solid. This transformation process is called solidification. While pure metals solidify at a particular temperature, alloys can solidify at different temperature. The process depends on the temperature gradient, under cooling, growth rate, and alloy constitution. This section also explores the concept of casting by which a liquid material is poured into a mold and left to solidify. All the diverse principles of solidification and casting have been carefully analyzed in this chapter.

Steel Solidification

Solidification is a phase transformation process in which liquid is transformed to solid when superheat and latent heat are removed. In this connection it is important to distinguish between solidification of pure metals and alloys. Pure metals solidify at constant temperature, whereas alloys solidify over a range of temperature. The following binary phase diagram illustrates the solidification in alloys:

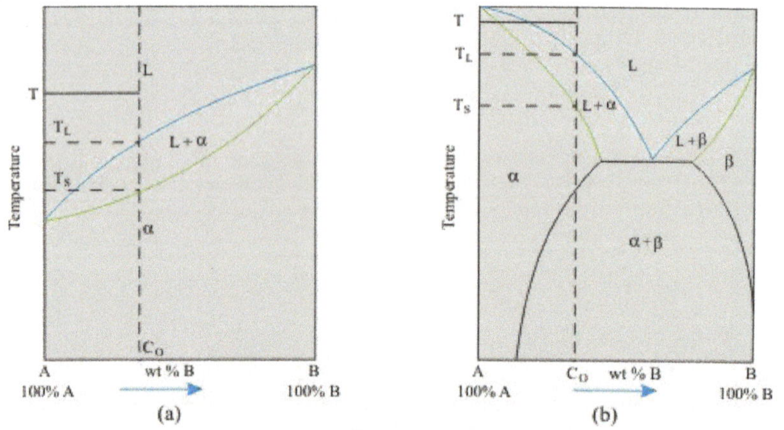

Equilibrium phase diagram of a binary system (a) there is a complete solubility of the components in liquid and solid and (b) complete solubility of the components in the liquid phase and partial solubility in the solid phase. α and β are solid solutions. Liquidus, solidus and solvus lines are shown by blue, green and black lines.

Vertical dotted line show the temperature changes that occurs during solidification of an alloy of composition CO.

Solidification Variables

Solidification behavior depends on parameters such as growth rate, temperature gradient, under cooling and alloy constitution. The temperature gradient (G) and growth rate (R) influence the solidification morphology and solidification substructure respectively.

Growth rate or solidification rate is the rate of advance of the solid/liquid interface into the liquid. The rate of movement of solidification front determines solute redistribution during solidification, scale of solidification substructure and the growth undercooling. Rapid solidification (faster movement of solid/liquid interface) minimizes the tendency of segregation of elements.

Temperature gradients both in solid and liquid are important. Thermal gradient in liquid is more critical as compared with solid and is strongly affected by mixing in liquid. Temperature gradient in solid is diffusion dependent.

Undercooling is the difference between the liquid temperature of the alloy of nominal composition and the actual temperature.

The total undercooling ΔT

$$\Delta T = \Delta T_{Th} + \Delta T_C + \Delta T_R + \Delta T_K$$

$\Delta T_C =$ Constitutional undercooling

$\Delta T_K =$ Kinetic undercooling,

$\Delta T_R =$ Undercooling due to curvature at solid/liquid interface.

$\Delta T_{Th} =$ = Thermal undercooling

ΔT_K and ΔT_R are usually negligible in casting. Thermal under cooling is required in solidification of pure metal where there is a significant nucleation barrier for liquid to solid transformation or when directional solidification is carried out at a lower rate where cellular structure forms. In the solidification of ingot and continuous casting where alloys are solidified, thermal undercooling is not of interest.

In solidification of alloy, the liquidus temperature of the melt is composition dependent. If the liquid composition deviates from the bulk composition, then the melt temperature may differ from the liquidus temperature of the overall alloy. The change in transformation temperature due to compositional effect is called constitution undercooling.

Equilibrium Solidification

Equilibrium solidification of an alloy of any composition occurs when concentration of the liquid and that of solid follows the liquidus and solidus line of the phase diagram. Though it is an idealized concept, equilibrium solidification helps understanding the compositional changes that occur during solidification. Solidification of an alloy of a composition begins when the temperature reaches to the liquidus line. Lowering of temperature changes the composition of elements both in liquid and solid phase. Composition of liquid and solid in the two phase region at any temperature is determined by a tie line. The tie line connects liquidus and solidus line at a temperature in the two phase region.

As equilibrium is assumed, at some temperature between T_L (liqiuidus temperature) and T_S (solidus temperature)

$$C_S f_S + C_L f_L = C_o$$

according to lever rule f_l and f_s are liquid and solid fraction respectively.

Non Equilibrium Solidification

Assumption of complete mixing does not prevail in actual solidification process. Diffusion in the solid is a slow process. Fluid flow in liquid will govern the concentration gradient of the solute in the liquid. Only faster mixing can minimize the concentration gradients, otherwise concentration gradients will be present in the liquid also. The presence of gradients in the solid and liquid during solidification makes the solidification process to deviate from the equilibrium. This means that in the actual solidification of alloy the concentration of solute in the liquid and solid rarely follows the liquidus and solidus lines of the respective phase diagram.

Consider solidification of an alloy of composition Co at temperature T in which there is no mixing in the solid but mixing in liquid is sufficiently large so as to have no composition gradient in the liquid. In other words liquid has a uniform composition at every stage of solidification. This can be illustrated by plotting concentration of solute vs distance both in the solid and liquid as shown in figure.

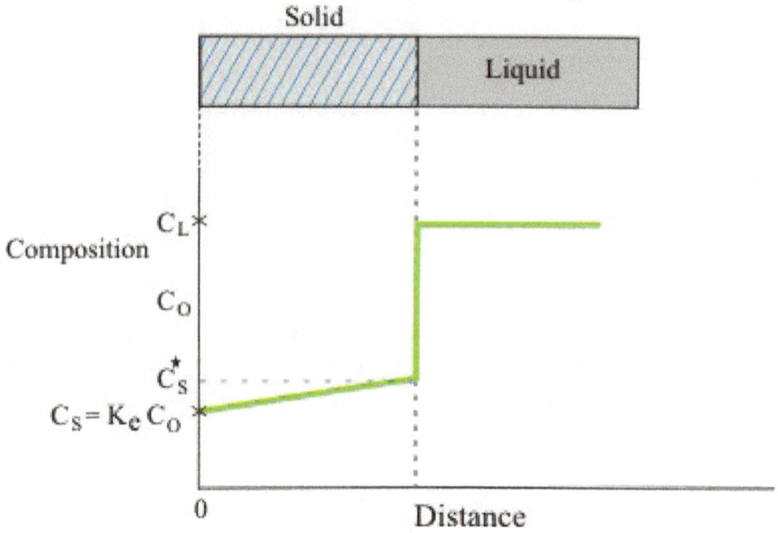

Concentration vs. distance profile in solid and liquid in one dimensional solidification.

At the onset of solidification $C_L = C_O$ and $C_S = K_e C_O$, where K_e is the partition coefficient, which is shown in the figure at X=0. As the solidification proceeds the composition of liquid follows the liquidus line but there will be gradient of solute since no mixing is assumed in solid. Figure shows the variation of concentration of solute in the solid by a green line. In the figure C_s^* is the interfacial concentration. As the solidification proceeds more and more solute will be rejected in the liquid but no gradient of the solute will be built-up in the liquid since complete mixing is assumed. Mass balance of solute when a small amount of solid df_s forms and causes solute to increase in liquid dC_L.

$$(C_L - C_S^*)df_s = f_L dC_L \quad (1)$$

$$f_s + f_L = 1 \;\; and \;\; K_e = \frac{C_S^*}{C_L} \quad (2)$$

$$By\ 1\ and\ 2\ and\ integrating\ the\ equate\ on\ C_L = K(1-f_s)^{K_e-1} \quad (3)$$

$$C_L = C_o(1-f_s)^{K_e-1} \quad (4)$$

Equation 4 is Sheil's equation. K_e is segregation coefficient. The role of partition coefficient of an element on solidification becomes clear from equation 4. At $K_e = 1$, $\frac{C_L}{C_O} = 1$ for all values of f_S. However, when K_e is < 1, $\frac{C_L}{C_O} > 1$ for all values of f_s. The value of K_e is given for some elements in iron in table.

Element	δ-iron	γ- iron
C	0.13	0.36
S	0.02	0.02
P	0.13	0.06
O	0.02	0.02
N	0.28	0.54
H	0.32	0.45
Mn	0.84	0.95
Ti	0.14	0.07

A more practical situation for solidification would be no mixing in either solid or in liquid. As the solid/liquid interface advances into the liquid, solute concentration builds up near the interface since no mixing in the liquid is assumed. There will be gradient of concentration in the liquid as illustrated in the figure.

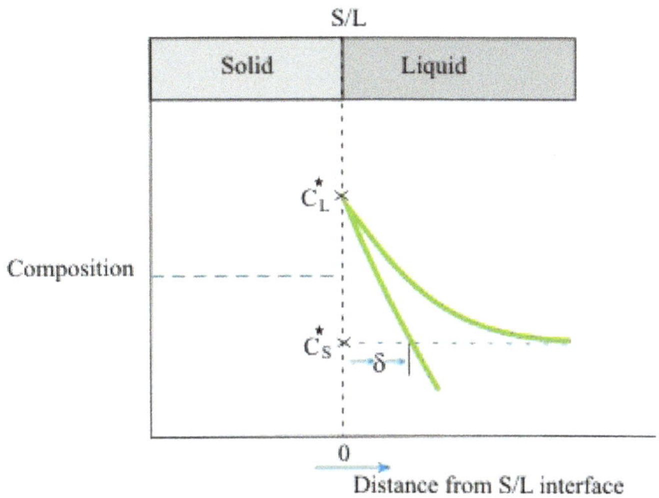

Concentration versus distance profile. In the figure δ is boundary layer thickness in which gradient of concentration persists

Microstructure Development

Nucleation and growth behavior, both determine development of microstructure in solidification of alloys. In alloys, nucleation of solid is not an issue since nucleating sites are present. Growth behavior is important.

Growth Behavior

The solute redistribution at the solid/liquid interface governs the stability of the solidification front. The solidification begins with the planar solid/liquid interface. During growth the morphological stability of the planer solid/liquid interface is governed by thermal and composition gradients at the interface.

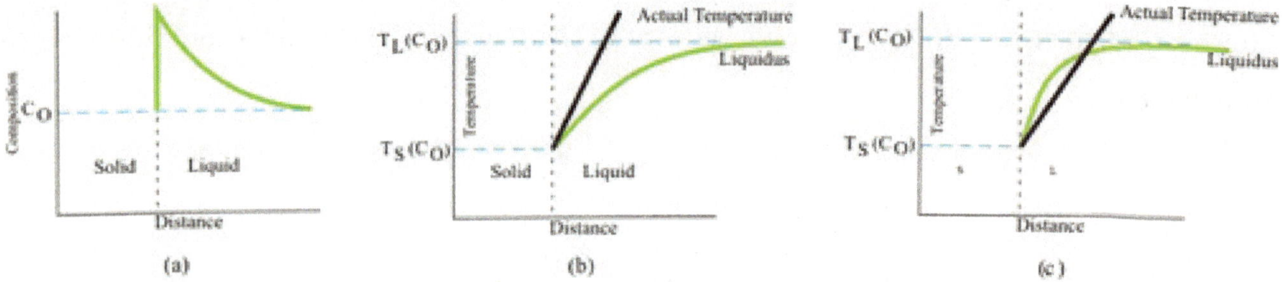

Constitutional super cooling during plane front alloy solidification, (a) concentration profile of the solute enriched layer ahead of the planar solidification front, (b) condition for a stable planar front solidification and (c) condition for unstable planar solid/liquid front.

In figure (a) the variation in composition as a function of distance from the planar solid/liquid interface is depicted. By referring to the phase diagram such as one shown in figure b, a curve of equilibrium liquidus temperature versus distance from the S/L interface corresponding to figure (a) can be determined and such a curve is shown in figure b and c. (It must be noted that the equilibrium liquidus temperature depends on the solute; as the solute segregates, the liquidus temperature will be lower than when there is no segregation.)

Two conditions are possible based on the actual temperature distribution in the liquid imposed by temperature gradient arising from the heat flow conditions:

Condition 1: Actual thermal gradient is steeper than equilibrium liquidus temperature at the interface due to the compositional gradient. Under this condition any perturbation developed at the solid/interface will remelt and the planar front will remain stable.

Condition 2: If, however, actual thermal gradient is smaller, as shown in figure c, then the liquid ahead of the interface is supercooled since the actual temperature ahead of the interface is lower than the equilibrium liquidus temperature. Any perturbation of the plane front can no longer melt but instead will grow. Under these conditions the planar growth is no longer stable, this is due to constitutional super cooling and is responsible for cellular or dendrite solidification.

The compositional gradient for steady state plane front growth can be determined by mass balance consideration at the planar interface to

$$\frac{G_L}{R} < \frac{\Delta T_O}{D_L} \qquad (5)$$

In equation 5 Δt_0 is the liquidus and solidus temperature difference at composition CO, D_L is the diffusion coefficient of solute in liquid, G_L is thermal gradient in the liquid. Several parameters affect the constitutional supercooling. Decrease in the imposed temperature gradient increases the tendency for planar solidification.

In addition to gradient, growth rate is also important. For a given gradient,(G), the growth rate (R) also plays an important role in determining solidification structure. Higher temperature gradient and smaller growth rate i.e. when G/R is large plane front solidification prevails. As G/R decreases, plane front can no longer be stable. Low G/R favors dendridic solidification. Solutes with a smaller partition coefficient create conditions for instable plane front solidification, since more solute will be rejected in the liquid. Thus as the extent of constitutional super cooling ahead of the solid/liquid interface increases, the morphology of interface changes from planar to cellular and to dendrite.

Casting

Molten metal before casting

Casting iron in a sand mold

Judenplatz Holocaust Memorial (Nameless Library), by Rachel Whiteread. Concrete cast of books on library shelves turned inside out.

Casting is a manufacturing process in which a liquid material is usually poured into a mold, which contains a hollow cavity of the desired shape, and then allowed to solidify. The solidified part is also known as a *casting*, which is ejected or broken out of the mold to complete the process. Casting materials are usually metals or various *cold setting* materials that cure after mixing two or more components together; examples are epoxy, concrete, plaster and clay. Casting is most often used for making complex shapes that would be otherwise difficult or uneconomical to make by other methods.

Casting is a 6000-year-old process. The oldest surviving casting is a copper frog from 3200 BC.

Types

Metal

In metalworking, metal is heated until it becomes liquid and is then poured into a mold. The mold is a hollow cavity that includes the desired shape, but the mold also includes runners and risers that enable the metal to fill the mold. The mold and the metal are then cooled until the metal solidifies. The solidified part (the casting) is then recovered from the mold. Subsequent operations remove excess material caused by the casting process (such as the runners and risers).

Plaster, Concrete, or Plastic Resin

Plaster and other chemical curing materials such as concrete and plastic resin may be cast using single-use *waste* molds as noted above, multiple-use 'piece' molds, or molds made of small rigid pieces or of flexible material such as latex rubber (which is in turn supported by an exterior mold). When casting plaster or concrete, the material surface is flat and lacks transparency. Often topical treatments are applied to the surface. For example, painting and etching can be used in a way that give the appearance of metal or stone. Alternatively, the material is altered in its initial casting process and may contain colored sand so as to give an appearance of stone. By casting concrete, rather than plaster, it is possible to create sculptures, fountains, or seating for outdoor use. A simulation of high-quality marble may be made using certain chemically-set plastic resins (for example epoxy or polyester which are thermosetting polymers) with powdered stone added for coloration, often with multiple colors worked in. The latter is a common means of making washstands, washstand tops and shower stalls, with the skilled working of multiple colors resulting in simulated staining patterns as is often found in natural marble or travertine.

Fettling

Raw castings often contain irregularities caused by seams and imperfections in the molds, as well as access ports for pouring material into the molds. The process of cutting, grinding, shaving or sanding away these unwanted bits is called "fettling". In modern times robotic processes have been developed to perform some of the more repetitive parts of the fettling process, but historically fettlers carried out this arduous work manually, and often in conditions dangerous to their health.

Fettling can add significantly to the cost of the resulting product, and designers of molds seek to minimize it through the shape of the mold, the material being cast, and sometimes by including decorative elements.

Casting process Simulation

Casting process simulation uses numerical methods to calculate cast component quality considering mold filling, solidification and cooling, and provides a quantitative prediction of casting mechanical properties, thermal stresses and distortion. Simulation accurately describes a cast component's quality up-front before production starts. The casting rigging can be designed with respect to the required component properties. This has benefits beyond a reduction in pre-production sampling, as the precise layout of the complete casting system also leads to energy, material, and tooling savings.

The software supports the user in component design, the determination of melting practice and casting methoding through to pattern and mold making, heat treatment, and finishing. This saves costs along the entire casting manufacturing route.

Casting process simulation was initially developed at universities starting from the early '70s, mainly in Europe and in the U.S., and is regarded as the most important innovation in casting technology over the last 50 years. Since the late '80s, commercial programs (such as AutoCAST and MAGMA) are available which make it possible for foundries to gain new insight into what is happening inside the mold or die during the casting process.

Ingot Casting

Molten steel from BOF/EAF is tapped into a teeming ladle. Deoxidizers, decarburizes and alloying elements if required, are added for the final finishing with respect to oxygen content and other elements in steel. The steel may be degassed either before or during casting. In the modern steel plants, steel is cast continuously. In several small scale plants, particularly those based on induction melting furnaces ingot casting is practiced.

Ingot casting is done in cast iron moulds having square, round or polygon cross section. Ingots with square cross section are used for rolling into billets, rails and other structural sections. Whereas, ingots with rectangular cross section (also known as slab), are used for rolling into flat products. Round ingots are used for tube making. Polygon ingots are used to produce tyres, wheels, etc. Typically an ingot weighing 5-20 tons for rolling, whereas few hundred to 300 tons for forging.

Ingot Mould Types

Cast iron is used to fabricate the mould. Thermal coefficient of cast iron is lower than steel as a result, steel on solidification contracts more than cast iron which makes detachment of ingot easier from the mold. Inner walls of the mould are coated by tar or fine carbon. The coated material decomposes during solidification which prevents sticking of solidified ingots with the inner walls of the mold.

Molds are essentially of two types:

1. Wide end up or narrow end down as shown in figure (a)
2. Narrow end up or big end down as shown in figure (b)

Wide end up moulds are used to produce forging ingots of killed plain carbon or alloy steels. Wide end up molds may have a solid bottom. Narrow end up molds are commonly used to produce rimming and semi-killed steel ingots. Narrow-end-up molds facilitates easy escape of rimming reaction product, CO.

Fully deoxidized or killed steel used for high quality forgings shrink on solidification and may lead to formation of pipe. Molds are generally provided with hot top which acts as reservoir to feed the metal and to avoid formation of pipe. Insulating and exothermic materials are put on the top ingot which ensures availability of hot metal towards the end of solidification.

Both bottom pouring and top pouring of steel are used in ingot casting.

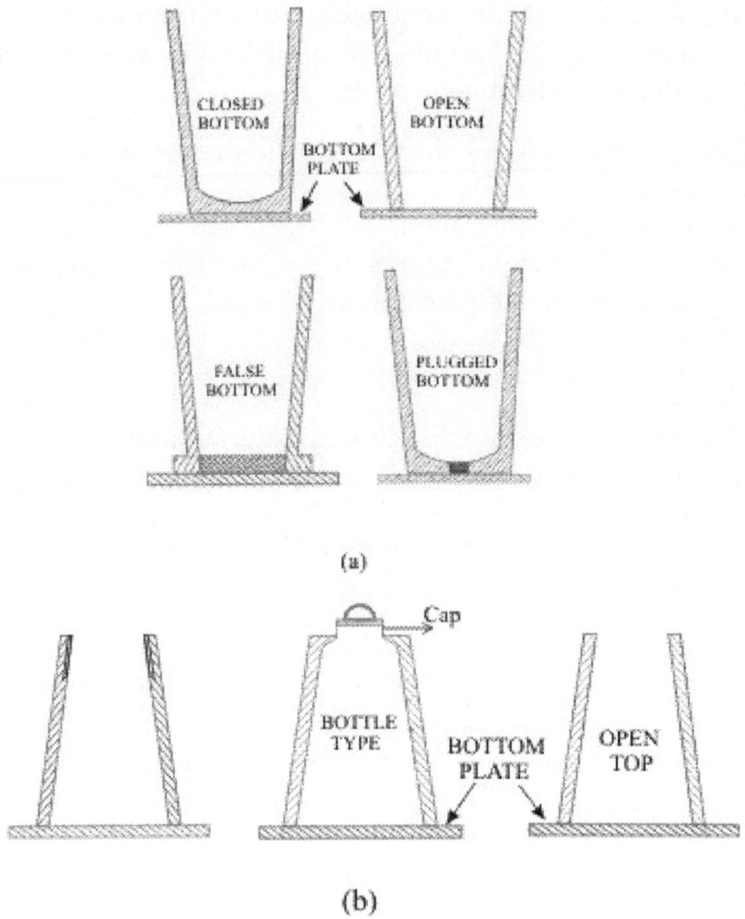

(a) wide end up moulds (b) Narrow end up moulds

Mechanism of Solidification

Molds are water cooled. Killed steel solidifies in the ingot form as follows:

i) Metal near the mould walls and bottom is chilled by the cold surfaces and a thin shell or skin is formed on the ingot surface. This surface has a fine equiaxed grains and the skin. The formation of skin results in decrease in rate of solidification.

ii) Due to expansion of mould through the heat transferred from the solidifying steel and contraction of solidified skin an air gap forms between the mould and the skin.Th is results in decrease in the heat transfer rate, because air gap has a high thermal resistance to heat flow

iii) The solidification front perpendicular to the mold faces moves inwards and towards the centre as a result columnar grains form next to the chill surface. The columnar crystals rarely extend to the centre of the mould.

iv) The central portion of the ingot solidifies as equi-axed grains of bigger size due to slow rate of solidification.

The above zones of solidification depend on the evolution of CO gas due to carbon and oxygen reaction. In semi killed steels, not all oxygen removed from steel. Oxygen content of steel is very low.

The necessary super saturation level of carbon and oxygen reaches towards the end of solidification. As a result the central zone of the equi-axed crystal is disturbed by way of formation of blow holes in the top middle potion of the ingot.

Solidification of rimming steels is controlled by evolution of CO during solidification. Rimming steels are not killed. The gas is evolved at the solid/liquid interface which stirs the molten steel during solidification. Stirring circulates molten steel which brings hot metal to the surface and solidification of steel at top is delayed. Columnar grain formation is prevented due to a more uniform temperature at interior of an ingot. This gives rise to rimming ingots in which gas is entrapped mechanically as blow holes.

Ingot Defects: Causes and Remedies

i) Pipe formation

Cause: Steel contracts on solidification. The volumetric shrinkage leads to formation of pipe. In killed steels pipe formation occurs toward the end of solidification. Figure shows primary and secondary pipe in narrow end up mould and (b) in wide end up mould while casting killed steel. Only primary pipe can be seen in wide end up mould.

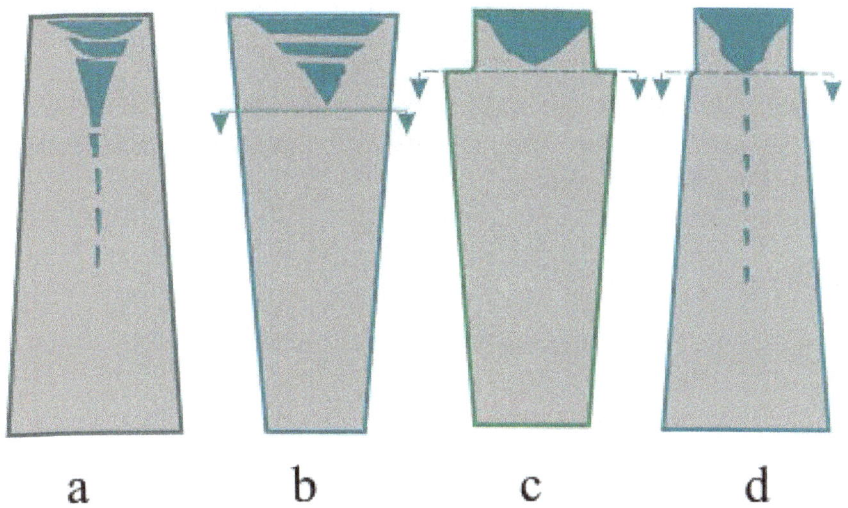

(a) Narrow end up mould showing long pipe in killed steel
(b) Wide end up mould showing pipe in killed steel
(c) Wide end up mould with hot top
(d) Narrow end up mold with hot top. Pipe is confined to the top.

Rimming and semi-finished steel show very less tendency for pipe formation

Wide end up moulds show smaller pipe as compared with narrow end up mould (in figure (a) longer pipe can be seen). The portion of ingot containing pipe has to be discarded which affects yields.

Remedy: use of hot top on the mold. The volume of the hot top is 10-15% higher than ingot volume. Pipe formation is restricted in the hot top which can be discarded. Use of exothermic materials in the hot top keeps the metal hot in the top portion and pipe formation can be avoided. Another method is to pour extra mass of metal.

ii) Blow holes

Cause: Evolution of gas during solidification of steel. Entrapment of gas produces blow holes in the ingot. Blow holes located inside the ingot can be welded during rolling. Rimming steels show blow holes due to rimming reaction between carbon and oxygen. The rimming reaction produces CO, which when is unable to escape during solidification, produces blow holes. Semi-killed steels also show tendency to blow hole formation.

Remedy: Control of gas evolution during solidification so that blow hole forms only within the ingot skin of adequate thickness.

Segregation: It is the difference in composition of steel within the ingot than some average composition. Segregation is due to

a) Difference in solubility of solute elements in liquid and solid steel i.e. partition coefficient of element in steel. Partition coefficient of solute (K) is defined as

$$K = \frac{\textit{Concentration of solute in solid}}{\textit{Concentration of the solute in liquid}}$$

The value of K ≤ 1 The solute elements whose K = 1 do not segregate. All elements whose K ≤ 1 tend to segregate.

b) Rate of solidification: faster rate of solidification avoids the elements to segregate. The initial chill layer of ingot has practically the same composition as that of liquid steel. Decrease in rate of solidification causes elements to segregate.

c) Larger size ingots are prone to segregation than smaller size ones. Larger size ingots require more time for solidification.

Remedy: soaking of ingots at high temperature can minimize segregation.

Non Metallic Inclusions

Non metallic inclusions are inorganic oxides, sulphides and nitrides formed by reaction between metal like Fe, Ti, Zr, Mn, Si, Al with non metallic elements like oxygen, nitrogen, sulphur etc. An inclusion is a mismatch with the steel matrix.

Fine size inclusions when distributed uniformly are not harmful. Non deformable inclusions like Al_2O_3 are undesirable.

Inclusion modification is the remedy to alleviate the harmful effect of inclusions on properties of steel.

Ingot Cracks

Surface cracks are formed due to friction between mold and ingot surface. The improper design of mold taper and corner radius cause surface cracks. Different types of cracks are:

Transverse cracks: They are parallel to the base of ingot and are formed due to longitudinal tension in the ingot skin. As the aspect ratio of the ingot increases, tendency to transverse crack formation increases. Longitudinal cracks are formed due to lateral tension in the skin. They are parallel to vertical axis of ingot. Alloy steels are more prone to longitudinal cracks than mild steels.

Sub- cutaneous cracks are internal fissures close to the surface. The cracks are formed due to thermal shocks.

Restriction cracks can be near the corner radius of the ingot.

Smooth corners of the mould and gradual curvature minimize restriction cracks.

Continuous Casting

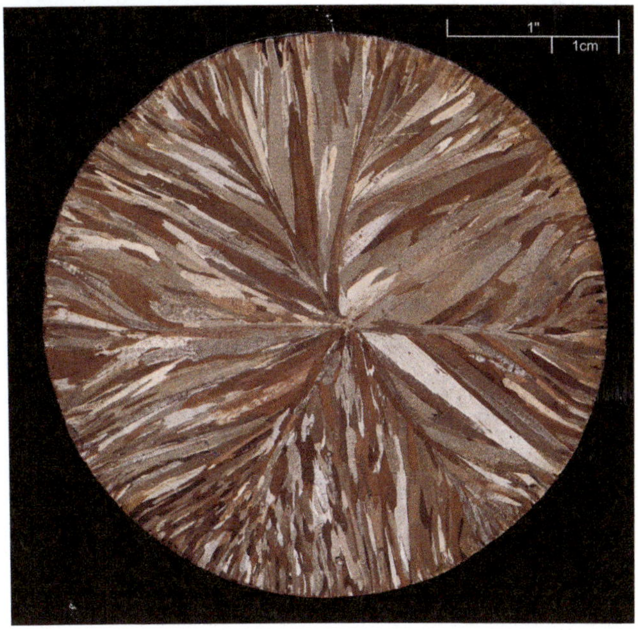

The macrostructure of continuously cast copper (99.95% pure), etched, Ø ≈ 83 mm.

Continuous casting, also called strand casting, is the process whereby molten metal is solidified into a "semifinished" billet, bloom, or slab for subsequent rolling in the finishing mills. Prior to the introduction of continuous casting in the 1950s, steel was poured into stationary molds to form ingots. Since then, "continuous casting" has evolved to achieve improved yield, quality, productivity and cost efficiency. It allows lower-cost production of metal sections with better quality, due to the inherently lower costs of continuous, standardised production of a product, as well as providing increased control over the process through automation. This process is used most frequently to cast steel (in terms of tonnage cast). Aluminium and copper are also continuously cast.

Sir Henry Bessemer, of Bessemer converter fame, received a patent in 1857 for casting metal between two counter-rotating rollers. The basic outline of this system has recently been implemented today in the casting of steel strip.

Equipment and Process

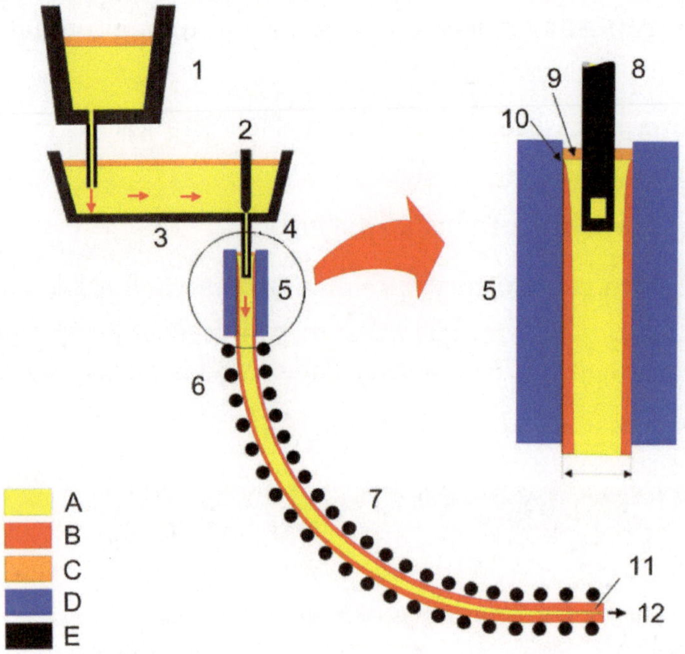

Continuous casting. 1: Ladle. 2: Stopper. 3: Tundish. 4: Shroud. 5: Mold. 6: Roll support. 7: Turning zone. 8: Shroud. 9: Bath level. 10: Meniscus. 11: Withdrawal unit. 12: Slab. A: Liquid metal. B: Solidified metal. C: Slag. D: Water-cooled copper plates. E: Refractory material.

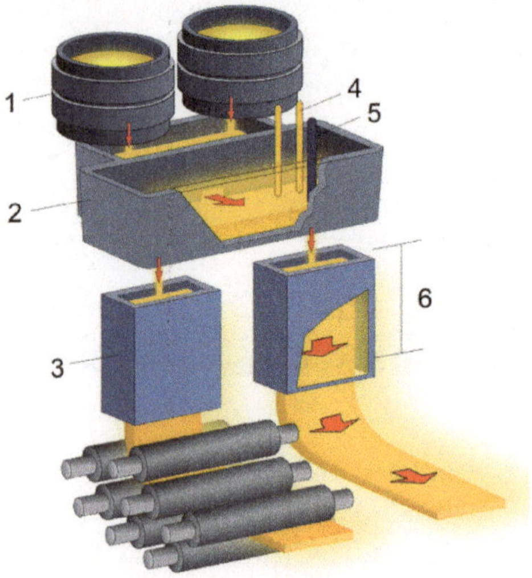

Continuous casting (Tundish and Mold). 1: Ladle. 2: Tundish. 3: Mold. 4: Plasma torch. 5: Stopper. 6: Straight zone.

Low-Carbon Steels

Molten metal is tapped into the ladle from furnaces. After undergoing any ladle treatments, such as alloying and degassing, and arriving at the correct temperature, the ladle is transported to the top of the casting machine. Usually the ladle sits in a slot on a rotating turret at the casting machine. One ladle is in the 'on-cast' position (feeding the casting machine) while the other is made ready in the 'off-cast' position, and is switched to the casting position when the first ladle is empty.

From the ladle, the hot metal is transferred via a refractory shroud (pipe) to a holding bath called a tundish. The tundish allows a reservoir of metal to feed the casting machine while ladles are switched, thus acting as a buffer of hot metal, as well as smoothing out flow, regulating metal feed to the molds and cleaning the metal.

Metal is drained from the tundish through another shroud into the top of an open-base copper mold. The depth of the mold can range from 0.5 to 2 metres (20 to 79 in), depending on the casting speed and section size. The mold is water-cooled to solidify the hot metal directly in contact with it; this is the *primary cooling* process. It also oscillates vertically (or in a near vertical curved path) to prevent the metal sticking to the mold walls. A lubricant (either powders that melt on contact with the metal, or liquids) is added to the metal in the mold to prevent sticking, and to trap any slag particles—including oxide particles or scale—that may be present in the metal and bring them to the top of the pool to form a floating layer of slag. The shroud is set so the hot metal exits it below the surface of the slag layer in the mold and is thus called a submerged entry nozzle (SEN). In some cases, shrouds may not be used between tundish and mold ('open-pour' casting); in this case, interchangeable metering nozzles in the base of the tundish direct the metal into the moulds. Some continuous casting layouts feed several molds from the same tundish.

In the mold, a thin shell of metal next to the mold walls solidifies before the middle section, now called a strand, exits the base of the mold into a spray chamber. The bulk of metal within the walls of the strand is still molten. The strand is immediately supported by closely spaced, water-cooled rollers which support the walls of the strand against the ferrostatic pressure (compare hydrostatic pressure) of the still-solidifying liquid within the strand. To increase the rate of solidification, the strand is sprayed with large amounts of water as it passes through the spray-chamber; this is the *secondary cooling* process. Final solidification of the strand may take place after the strand has exited the spray-chamber.

It is here that the design of continuous casting machines may vary. This describes a 'curved apron' casting machine; vertical configurations are also used. In a curved apron casting machine, the strand exits the mold vertically (or on a near vertical curved path) and as it travels through the spray-chamber, the rollers gradually curve the strand towards the horizontal. In a vertical casting machine, the strand stays vertical as it passes through the spray-chamber. Molds in a curved apron casting machine can be straight or curved, depending on the basic design of the machine.

In a true horizontal casting machine, the mold axis is horizontal and the flow of steel is horizontal from liquid to thin shell to solid (no bending). In this type of machine, either strand or mold oscillation is used to prevent sticking in the mold.

After exiting the spray-chamber, the strand passes through straightening rolls (if cast on other than a vertical machine) and withdrawal rolls. There may be a hot rolling stand after withdrawal to take advantage of the metal's hot condition to pre-shape the final strand. Finally, the strand is cut into predetermined lengths by mechanical shears or by travelling oxyacetylene torches, is marked for identification, and is taken either to a stockpile or to the next forming process.

In many cases the strand may continue through additional rollers and other mechanisms which may flatten, roll or extrude the metal into its final shape.

Casting Machines for Aluminium and Copper

continuous hot vertical casting in process (aluminum)

molten aluminum pours into this casting die (top view of die)

bottom end of casting die

the resulting Aluminum blanks (after cutting to size)

Aluminium and copper can be cast horizontally and can be more easily cast into near net shape, especially strip, due to their lower melting temperatures.

Range of Continuously Cast Sections

- Casting machines are designated to be billet, bloom or slab casters.
- Slab casters tend to cast sections that are much wider than thick:
    - Conventional slabs lie in the range 100–1600 mm wide by 180–250 mm thick and up to 12 m long with conventional casting speeds of up to 1.4 m/minute; however slab widths and casting speeds are currently increasing.
    - Wider slabs are available up to 3250×150 mm
    - Thick slabs are available up to 2200×450 mm at a specific steel facility, generically ranging typically from 200mm to 300mm
    - Thin slabs (low-carbon steel): 1680×50 mm at a specific facility, generically ranging from 40mm to 110mm thick depending on an individual machine's design

- Conventional bloom casters cast sections above 200×200 mm. The bloom length can vary from 4 to 10 m
- Billet casters cast smaller section sizes, such as below 200 mm square, with lengths up to 12 m long. Cast speeds can reach up to 4 m/minute.
- Rounds: either 500 mm or 140 mm in diameter
- Conventional beam blanks: look similar to I-beams in cross-section; 1048×450 mm or 438×381 mm overall
- Near net shape beam blanks: 850×250 mm overall
- Strip: 2–5 mm thick by 760–1330 mm wide

Startup, Control of the Process and Problems

Starting a continuous casting machine involves placing a dummy bar (essentially a curved metal beam) up through the spray chamber to close off the base of the mould. Metal is poured into the mould and withdrawn with the dummy bar once it solidifies. It is extremely important that the metal supply afterwards be guaranteed to avoid unnecessary shutdowns and restarts, known as 'turnarounds'. Each time the caster stops and restarts, a new tundish is required, as any uncast metal in the tundish cannot be drained and instead freezes into a 'skull'. Avoiding turnarounds requires the meltshop, including ladle furnaces (if any) to keep tight control on the temperature of the metal, which can vary dramatically with alloying additions, slag cover and deslagging, and the preheating of the ladle before it accepts metal, among other parameters. However, the cast rate may be lowered by reducing the amount of metal in the tundish (although this can increase wear on the tundish), or if the caster has multiple strands, one or more strands may be shut down to accommodate upstream delays. Turnarounds may be scheduled into a production sequence if the tundish temperature becomes too high after a certain number of heats or the service lifetime of a non-replaceable component (i.e., the submerged entry nozzle (SEN) in a thin-slab casting machine) is reached.

Many continuous casting operations are now fully computer-controlled. Several electromagnetic, thermal, or radiation sensors at the ladle shroud, tundish and mould sense the metal level or weight, flow rate and temperature of the hot metal, and the programmable logic controller (PLC) can set the rate of strand withdrawal via speed control of the withdrawal rolls. The flow of metal into the moulds can be controlled via three methods:

- By stopper rods that descend through the tundish,
- By slide gates at the top of the mould shrouds,
- If the metal is open-poured, then the metal flow into the moulds is controlled solely by the internal diameter of the metering nozzles. These nozzles are usually interchangeable.

Overall casting speed can be adjusted by altering the amount of metal in the tundish, via the ladle slide gate. The PLC can also set the mould oscillation rate and the rate of mould powder feed, as well as the flow of water in the cooling sprays within the strand. Computer control also allows vital casting data to be transmitted to other manufacturing centres (particularly the steelmaking furnaces), allowing their work rates to be adjusted to avoid 'overflow' or 'underrun' of product.

Contamination by Oxygen

While the large amount of automation helps produce castings with no shrinkage and little segregation, continuous casting is of no use if the metal is not clean beforehand, or becomes 'dirty' during the casting process. One of the main methods through which hot metal may become dirty is by oxidation, which occurs rapidly at molten metal temperatures (up to 1700 °C for steel); inclusions of gas, slag or undissolved alloys may also be present. To prevent oxidation, the metal is isolated from the atmosphere as much as possible. To achieve this, exposed liquid metal surfaces are covered – by the shrouds, or in the case of the ladle, tundish and mould, by synthetic slag. In the tundish, any inclusions that are less dense than the liquid metal – gas bubbles, other slag or oxides, or undissolved alloys – may also float to the surface and be trapped in the slag layer. While the tundish and mold fill for the first time at the start of a casting run, the liquid is badly contaminated with oxygen and the first items produced are typically quarantined or diverted to customers who do not require top-quality material.

Breakouts

A major problem that may occur in continuous casting is *breakout* of the liquid metal: for whatever reason, the solid shell of the strand breaks and allows the still-molten metal contained within to spill out and foul the machine. In most industrial environments this event is very costly as it leads to a shutdown of the strand and typically requires an extended turnaround involving removal of the spilled material from within the strand equipment and/or replacement of damaged machinery. A breakout is usually due to the shell wall being too thin to support the liquid column above it, a condition which has several root causes often related to heat management. Improper cooling water flow to the mould or the strand cooling sprays may lead to inadequate heat removal from the solidifying metal, causing the solid shell to thicken too slowly. If the metal withdrawal rate is too fast, the shell may not have time to solidify to the required thickness even with enhanced cooling sprays. Similarly, the incoming liquid metal may be too hot and the final solidification may occur further down the strand at a later point than expected; if this point is below the straightening rolls, the shell may break from stresses applied during straightening. A breakout can also occur as a result of physical irregularities or damage to the shell occurring within the mould during the initial seconds of solidification. Excessive turbulence within the mold may cause an irregular shell pattern that grows abnormally or it may entrap slag droplets within the shell which reduces the wall strength. A common occurrence is for the shell to stick to the mould's surface and tear; modern instrumented molds and computer control systems typically detect this and slow the caster down temporarily to let the wall refreeze and heal while it is still supported in the mould. Should the tear occur near the exit of the mould or be of unexpected severity, the shell may still fail in a breakout once it emerges from the mould wall. If the incoming metal is severely overheated, it may be preferable to stop the caster than to risk a breakout. Additionally, lead contamination of the metal (caused by counterweights or lead-acid batteries in the initial steel charge) can form a thin film between the mould wall and the steel, inhibiting heat removal and shell growth and increasing the risk of breakouts.

Other Considerations

Another problem that may occur is a *carbon boil* – oxygen dissolved in the steel reacts with also-present carbon to generate bubbles of carbon monoxide. As the term *boil* suggests, this reaction is extremely fast and violent, generating large amounts of hot gas, and is especially dangerous if it occurs in the confined spaces of a casting machine. Oxygen can be removed by "killing" it through the addition of silicon or aluminium to the steel, which reacts to form silicon oxide (silica) or aluminium oxide (alumina). However, too much alumina in the steel will clog the casting nozzles and cause the steel to 'choke off'.

Computational fluid dynamics and other fluid flow techniques are being used extensively in the design of new continuous casting operations, especially in the tundish, to ensure that inclusions and turbulence are removed from the hot metal, yet ensure that all the metal reaches the mould before it cools too much. Slight adjustments to the flow conditions within the tundish or the mould can mean the difference between high and low rejection rates of the product.

Starter Bar

The starter bar, also called a dummy bar, has a free end portion which is flexible for storage and a substantially rigid portion at the end which plugs the mold. The starter bar is constructed in discrete blocks secured to one side of a planar spine provided in segments and arranged end to end. Adjustable spacers in the form of tapered blocks are disposed between the blocks of the bar to allow the starter bar to be self-supporting in a curved configuration corresponding to the casting path. A more flexible spine in the end portion of the starter bar allows the starter bar to be curved to a tighter radius than that of the casting path while the blocks fan out in an unsupported configuration. A storage ramp is provided to support the flexible end in the stored position. Before a cast is started, the starter bars are fed through the caster (in reverse direction of casting) using hydraulic actuators. Once fed all the way to the bottom of the mold, the process of packing the mold can continue to ensure a smooth start up.

Direct Strip Casting

Direct strip casting is a continuous casting process for producing metallic sheet directly from the molten state that minimizes the need for substantial secondary processing. For low-carbon sheet steels, this is a relatively new process which has only achieved commercial industrialization within the last decade.

Twin-belt Continuous Casting

Twin-belt continuous casting is a continuous casting process that produces high volume continuous metal bar or strip of constant rectangular cross section. Twin-belt continuous casting employs a moving mold consisting of parallel carbon-steel belts held in tension as top and bottom casting surfaces. Chains of rectangular steel or copper blocks moving with the belts and spaced according to the desired cast width form the sides of the mold.

Molten metal is introduced into the twin-belt continuous casting machine from a tundish through a nozzle placed between the casting belts. The metal is cooled by direct contact with the belts which are in turn cooled by high pressure recirculating water. Various coatings can be applied to the belt casting surfaces to provide required mold interface characteristics and prevent adhesion.

The cast metal from the twin-belt continuous casting machine is synchronized with, and directly fed into, a hot rolling mill. Combining the casting and rolling operations can result in significant energy and cost savings over other casting processes which incorporate intermediate cast and re-heat steps.

Metals Cast on Twin-Belt Continuous Casting Machines: Copper (Bar, Strip, Anode), Aluminum(Strip), Zinc (Strip), Lead (Strip)

Production rates and speeds: Twin-belt continuous casting rates range up to 60 tons per hour at speeds up to 14 meters per minute.

Twin-belt continuous casting is a near net shape casting process, which significantly reduces the need for secondary rolling or forming operations. For example, when casting copper anode plate the cast slab is not rolled but rather sheared directly into distinct anode plates.

The cooling belts are typically made of low carbon steel and are held under tension within the casting machine to ensure flatness and accuracy. As a "cold" belt enters the mold region, it is heated in the cast zone and is subject to powerful forces caused by thermal expansion. When casting wide strip, these forces must be controlled to eliminate buckling and reduce thermal distortion of the belt at the mold entrance. These forces can be controlled by preheating the belts before mold entry, or by magnetically stabilizing them once they have entered the mold.

Belt Preheating: For wide strip casting, a belt preheating system can be used to bring the belt up to 150 °C or higher immediately prior to entering the casting mold, reducing the effects of cold framing. Induction heating coils can be used across the width to preheat each belt. In addition to preventing thermal distortion, the high preheat temperature serves to eliminate any moisture present on the belt surface.

Magnetic Stabilization: When casting wide strip, the tendency of localized thermal distortion can be resisted by the use of high-strength, magnetic belt back-up support rolls within the mold region. The moving belt is held against the support rolls by magnetized rotating fins maintaining the belt in a flat plane.

Within the twin-belt continuous casting machine, molten metal progressively solidifies on the mold surfaces as it moves thru the mold region, with a sump of molten metal present between the solidifying outer surfaces. Belt coatings, texture, and gas layer modifications are used to fine tune the heat transfer rate from the cast metal to the belt. Full thickness solidification can occur as early as 30% of the way through the mold for thin strip, or up to 2 m beyond the mold exit for large bar where exit water spray cooling and roller support are required.

Closed Pool Feeding: When casting certain metals such as aluminum, a fully closed pool "injection" metal feeding system can be employed. Here, the metal is introduced under slight pressure into the closed mold cavity. Metal flow is controlled by maintaining a preset level in the tundish. The feed snout, or nozzle, is typically made from a ceramic material which is thermally stable and permeable to gases being released from the flowing metal.

Open Pool Feeding: When casting other metals, such as copper, zinc and lead, an open pool feeding system is often used. In this case, the upper belt pulley is offset downstream from the bottom pulley. Metal flows through an open trough or tundish into a standing pool of molten metal formed at the convergence of the belts. Shrouding gases may be employed to protect against oxidation.

Mold Tapering: The twin-belt casting machine differs from other moving mold casting machines in that all four mold surfaces are independent. This allows the mold surfaces to be tapered to remain in contact with the cast product as it shrinks. The high velocity cooling water, which is continuously applied to the backside of the belt, impinges on the belt and creates a force on the belt. This force acts to press the belt against the surface of the strip or slab as it shrinks, keeping the belt in close contact with the cast product throughout the mold. Each side of the mold is formed by an endless chain of dam blocks,which are held against the cast strip by adjustable spring-loaded guides.

Molten Metal Level Control: To accommodate high casting speeds and maintain as high a pool level as possible, non-contact electromagnetic metal level indicators can be used to sense the pool level in the casting machine.

Aluminum or copper strip casting: Commercial twin-belt continuous strip casting machines are capable of producing as-cast dimensions from 10–35 mm thick, and up to 2035 mm wide. After being directly fed into a hot rolling mill, the as-cast strip is typically rolled down to 1–3 mm thickness strip.

Copper bar casting: As-cast dimensions range from 35–75 mm thick, and from50-150 mm wide. After being directly fed into a hot rolling mill, the as-cast bar is typically rolled into 8 mm diameter rod to be used for wire drawing.

Copper anode casting: Special dam blocks which contain anode lug molds and a traveling hydraulic shear are added to the twin-belt casting machine to continuously cast net shape copper anodes. Anode width of approximately 1 meter (excluding lugs) and thicknesses from 16 mm to 45 mm. The primary advantage of this process is uniformity of the as-cast anode in terms of size and surface quality. Anodes cast using this process do not require additional preparation after casting.

Mold Length: The mold length ranges from approximately 2000 mm for strip casting machines and up to 3700 mm for copper bar casting machines.

In the continuous casting, molten steel is poured from the tundish in the water cooled mold and partially solidified bloom/billet or slab (hereafter called strand) is withdrawn from the bottom of the mold into water spray so that solidified bloom/billet or slab is produced constantly and continuously. Continuous casting is widely adopted by steelmakers. The advantages of continuous casting over ingot casting are:

- Quality of the cast product is better
- No need to have slabbing/blooming or billet mill as required when ingot casting is used.
- Higher extent of automation is possible

- Width of the slab can be adjusted with the downstream strip mill.
- Continuously cast products show less segregation.
- Hot direct charging of the cast product for rolling is possible which leads to energy saving.

How Casting is Done Continuously?

The essential components of a continuous casting machine are tundish, water cooled mold, water spray and torch cutters. Tundish, mold and water spray are arranged such that molten stream is poured from tundish to mold and solidified strand (billet/bloom/billet) is produced continuously. The required length of the strand is cut by torch cutter. In figure, the arrangement of tundish, mold and water spray is shown.

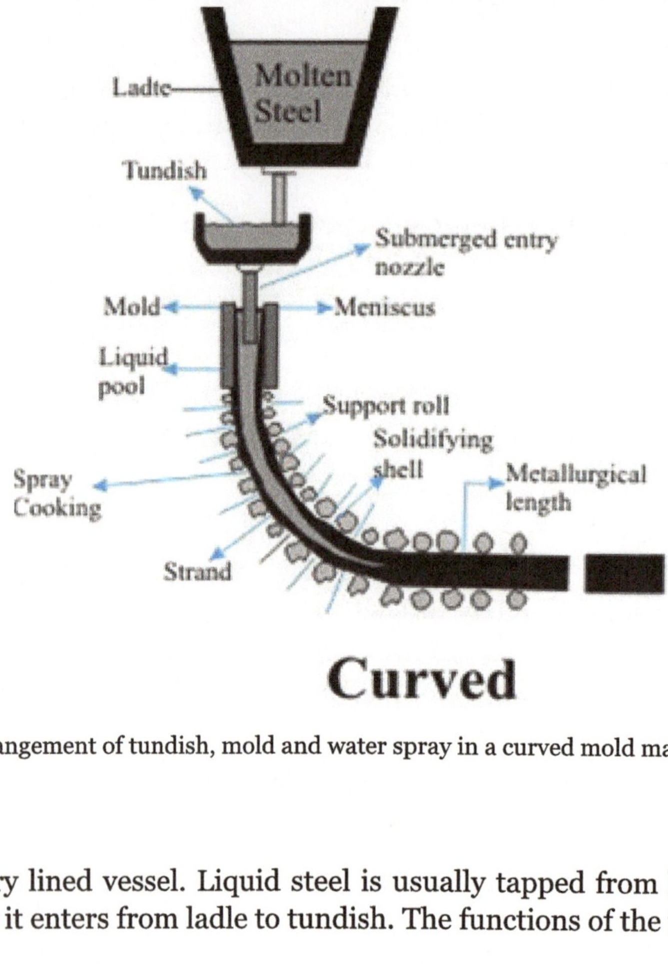

Arrangement of tundish, mold and water spray in a curved mold machine

Tundish

Tundish is a refractory lined vessel. Liquid steel is usually tapped from ladle into tundish. The stream is shrouded as it enters from ladle to tundish. The functions of the tundish are:

Reservoir of Molten Steel

Tundish acts as a reservoir for molten steel. It supplies molten steel in presence of a slag cover to all continuous casting molds constantly and continuously at constant steel flow rate. The flow rate is maintained constant by maintaining a constant steel bath height in the tundish through teeming of molten steel from the ladle. The number of mold is either one or more than one. Normally bloom and billet casting machines are multi-strand i.e. number of molds are either 4 or 6 or 8. Slab casters usually have either single or two molds. During sequence casting and ladle change- over periods, tundish supplies molten steel to the molds.

Distributor

Tundish distributes molten steel to different molds of the continuous casting machine at constant flow rant and superheat which is required for stand similarly with reference to solidification microstructure. Control of superheat is required in all the moulds to reduce break-out. Location of ladles stream in the tundish is important. It may be located symmetric or asymmetric to the centre of the tundish depending on the number of mold. For single strand machines, molten stream enters from one side and exits the other side of the tundish. In multi-strand tundishes, ladle stream is either at the centre of the tundish or displaced to the width side of the tundish.

Inclusion Removal

Tundish helps to remove inclusions during the process of continuous casting. For this purpose liquid steel flow in the tundish is modified by inserting dams, weirs, slotted dams etc. The whole idea is to utilize the residence time available before steel leaves the tundish. For example, if capacity of tundish is 40 tons and casting speed is 5 tons/min, then the average residence time of molten steel in the tundish is 8 minutes. During this average residence time., inclusion removal can be exercised .For this purpose flow of steel melt in the tundish has to be modified so as to accelerate the inclusion removal. The Inclusion removal is a two step step unit operation, namely floatation and absorption by a flux added on the surface of the tundish. Flux is usually rice husk, or fly ash or some synthetic powder. Figure shows a tundish with flow control devices employed to facilitate inclusion floatation.

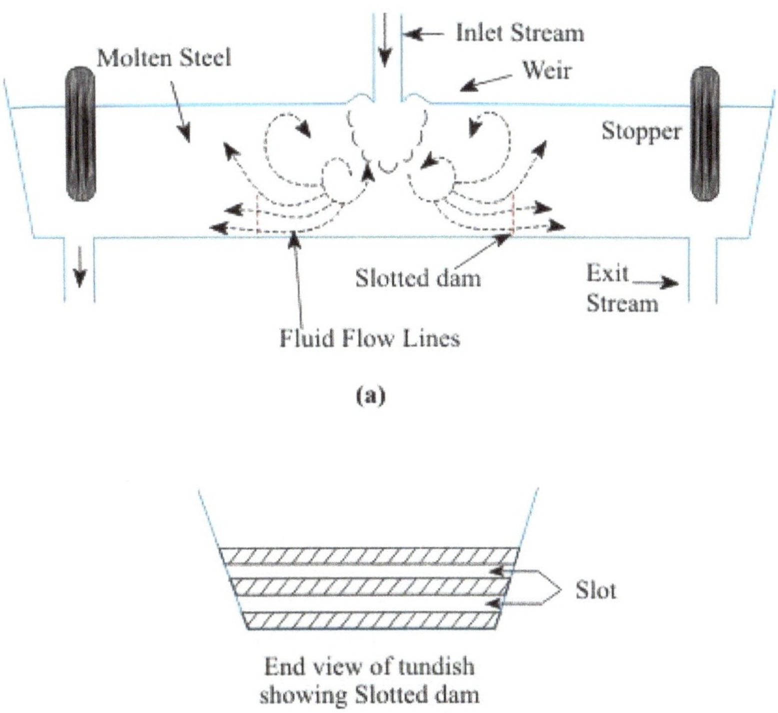

Tundish with flow control device, namely weir and slotted dam

Mold

Mold is the heart of continuous casting. Molten stream enters from the tundish through the submerged nozzle into water cooled mold in presence of flux. Solidification of steel begins in the mold. The casting powder is added onto the top of molten steel in the mold. It melts and penetrates between the surface of mold and the solidifying strand to minimize friction as shown in figure. Control of height of molten steel in the mould is crucial for the success of the continuous casting machine. The solidification begins from the meniscus of steel level in the mould. Mold level sensors are used to control the meniscus level in the mould.

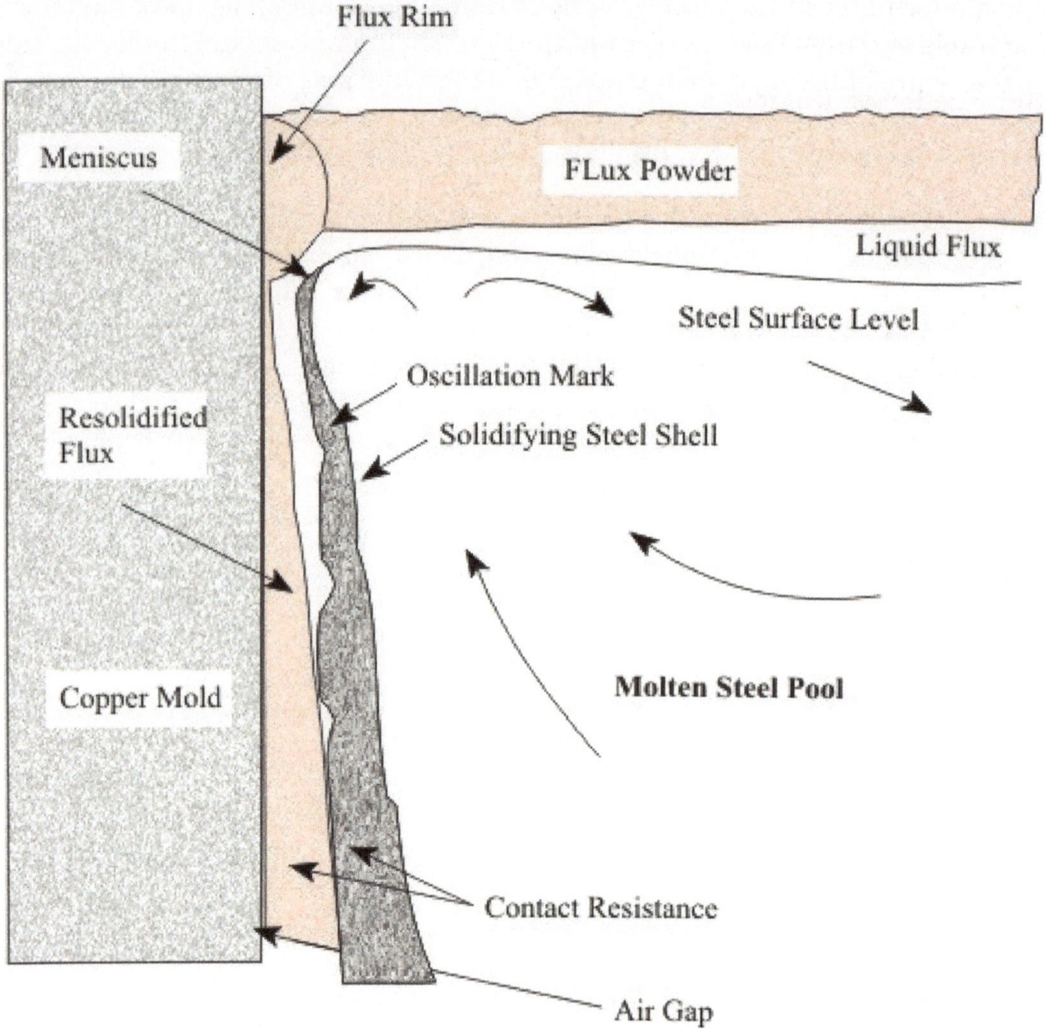

Role of flux in continuous casting mold

As seen in the figure, flux melts and enters into the gap between mold surface and solidified strand. Molds are made of copper alloys. Small amounts of alloying elements are added to increase the strength. Mold is tapered to reduce the air gap formation. Taper is typically 1% of the mold length. For $100\,mm \times 100\,mm$ cross section of mold the taper is about 1mm for 1m long mold. The cross section of the mold is the cross section of the slab/bloom/billet. Length of the mold is around $0.75-1.4\,m$ and is more for large cross sections. Mold cross section decreases gradually from top to bottom. Mould extracts around 10% of the total heat.

The mold is oscillated up and down to withdraw the partially solidified strand (strand is either billet or bloom or slab). The oscillated frequency can be varied. At Tata steel slab caster frequency is varied in between 0 and 250 cycles/min and the stroke length from 0 to 12mm.

Steel level in mould is controlled, that is the meniscus for smooth caster operation. Sensors are used to control the meniscus level.

The Functions of Mold Flux are

- Inclusion absorption capability.
- Prevention of oxidation.
- Minimization of heat losses.
- Flux on melting enters into the air gap and provides lubrication.

For the above Functions the Flux should have the following Properties.

- Low viscosity
- Low liquidus temperature
- Melting rate of flux must match with the speed of the continuous casting.

Mass flow rate of flux can be calculated by

$$\dot{m} = \rho\left(\frac{U_s\delta}{2} + \frac{\rho g \delta^s}{12\mu}\right)$$

$\dot{m} =$ Powder feed rate $kg/sm, U_S$ casting speed $m/s, \delta$ is boundary layer thickness, ρ is density of flux, μ is viscosity of slag (kg/ms).

Consider slab casting speed $0.05ms^{-1}, \mu = 3\times10^{-1}\frac{kg}{ms}, \delta = 0.1mm, \rho = 3000kgm^{-3}$

$$\dot{m} = 0.1\frac{kg}{sm}$$

For a mold of length $1m, \dot{m} = 6\,kg/\min$

Typically the range of composition for mold fluxes are.

CaO	25-45%	Na_2O	1-20%	BaO	0-10%
SiO_2	20-50%	K_2O	1-5%	Li_2O	0-4%
Al_2O_3	0-10%	FeO	0-6%	B_2O_3	0-10%
TiO_2	0-5%	MgO	0-10%	F	4-10%
C	1-25%	MnO	0-10%		

Design of Mold Flux

There are specific requirements of mould flux for specific grade of steel. For example,low carbon aluminum killed steel requires flux which can absorb Al_2O_3 inclusion without an adverse effect on viscosity. A lower viscosity helps the flux provide sufficient lubrication at higher casting speed.

Medium carbon grades ($0.08\%C\ to\ 0.18\%$) are prone to cracking. High solidification temperature of flux reduces heat through mold. For adequate lubrication low viscosity of the flux is required.

High carbon grades too require flux of low viscosity and melting point.

Ultra low carbon steels ($C < 0.005\%$) requires flux which can absorb non metallic inclusions, improve insulation, provided good lubrication, stable properties and minimal slag entrapment.

Table gives effect of chemical composition on mold flux properties.

Increase In	Viscosity	Solidification Temperature	Melting Point
CaO	Decrease	Increase	Increase
SiO_2	Increase	Decrease	Decrease
CaO/SiO_2	Decrease	Increase	Increase
Al_2O_3	Increase	Decrease	Increase
Na_2O	Decrease	Decrease	Decrease
F	Decrease	Increase	Decrease
Fe_2O_3	Decrease	Decrease	Decrease
MnO	Decrease	Decrease	Decrease
MgO	Decrease	Decrease	Decrease
B_2O_3	Decrease	Decrease	Decrease
BaO	Decrease	Decrease	Decrease
Li_2O	Decrease	Decrease	Decrease
TiO_2	No change	Increase	Increase
K_2O	Decrease	Decrease	Decrease

The table can be read as for example increase in CaO decreases viscosity but increases solidification and melting temperature of the flux. Similarly the effects on other constituents on the viscosity and solidification/melting temperature can be understood.

Secondary Cooling

Below the mold partially solidification strand is water sprayed to complete the solidification. Number of primary parameters which influence the rate of heat extraction are:

- Water drop flux
- Mean drop size

- Droplet velocity hitting the strand surface
- Wetting effects.

Spray cooling essentially involves boiling heat transfer. A water vapour blanket forms on the strand surface which prevents direct contact of water droplets with the strand surface. Velocity of droplets should be such that droplet can penetrate the vapour layer so that droplets can wet the surface and cools the surface.

In secondary cooling, number of nozzles is distributed over the surface of the moving strand. Overlapping of spray may occur. Distance between nozzles is important.

Heat Transfer in Continuous Casting

Heat transfer in continuous casting takes place in mold and in secondary cooling by a combination of conduction, convection and radiation. Figure shows heat transfer in the mold and secondary cooling.

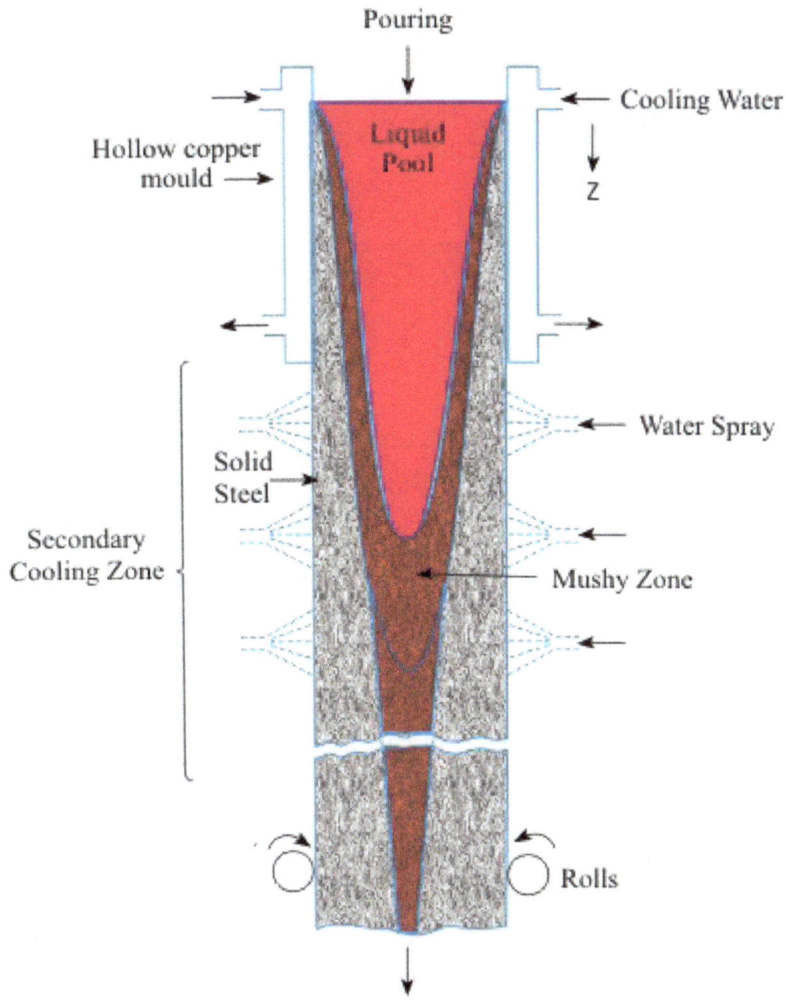

Heat transfer in the mold and secondary cooling zone and the formation of solid shell. Mushy zone and liquid core can also be seen

In the mold air gap formation influences heat transfer. The higher heat flux in mould can lead to higher casting speeds. Heat flux depends on:

- Composition of steel.
- Mould taper.
- Type of lubricant.
- Type of mould straight or curve.
- Casting speed.

The major requirements for secondary cooling are:

- Partially solidified strand must have shell sufficiently strong at the exit of the mold to avoid breakout due to liquid pressure.
- The liquid core should be bowl shaped.
- Solidification must complete before the withdrawl roll.

Casting speed i.e. rate of linear movement of strand/ minute from the mould is important. Casting speed must match with the rate of solidification. Slabs are cast within the speed ranging from 1.5/ min to 2.5m/min.

The intensity of heat extraction by water spray in secondary cooling is

$$h = \frac{heat\ flux}{T_S - T_W}$$

h is heat transfer coefficient $(W/m^2.s), T_s$ is surface and T_w water temperature. The heat transfer coefficient h depends on water flow rate. In secondary cooling solidification must be complete. Some of the issues are:

- Water spray must be distributed uniformly on the moving strand so that reheating of the strand does not occur. Non- uniform cooling leads to generation of thermal stresses on the surface and surface cracks may appear.
- Outer surface temperature should be greater than 850°C to avoid volumetric expansion accompanying due to transformation of austenite to ferrite.

Mist spray cooling i.e. mixture of air+ water provides more uniform cooling. Here high pressure air+ water mixture is sprayed on the metal surfaces. Some advantages are:

- Uniform cooling
- Less water requirement
- Reduced surface cracking

Products and Casting Defects

Presently killed steels are cast continuously into slab for flat products and bloom and billet for structural products.

Defects in continuous casting originate from several factors like mould oscillation, mould flux, segregation coefficient of solute elements; phase transformation etc. In the following, a brief presentation is given on defect formation.

Defects		
• Internal	• Surface	shape
• Midway cracks	• Longitudinal mid face and corner cracks	Rhomboidity
• Triple point cracks	• Transverse mid face and corner cracks.	Longitudinal depression ovality
• Center line cracks	• Deep oscillation masks	
• Diagonal cracks		
• Center segregation and porosity		
• Casting flux inclusion.		
• Blow holes		

Cracks are originated in the cast product due to mechanical and thermal stresses. Material factors are also responsible.

Mechanical Stresses are Created due to

- Friction.
- Ferro static pressure.
- Bending and straightening operation.
- Roll pressure.

Mechanical Stresses can be Reduced by Improving Mold Practices Like

- Controlling powder feed rate
- Resonance in mold
- More accurate strand guidance
- Casting powder

Thermal stresses are due to non-uniform cooling in the secondary zone. Controlling water flux impinging the surface of the strand and minimizing reheating of strand can alleviate thermal stresses. Also air +water mist spray provides more uniform cooling.

Material factors are related to $\delta-\gamma$ transformation. High S and low Mn/s ratio cause mid face longitudinal cracks. Control of inclusion is also important.

Electromagnetic Stirring (EMS)

EMS is an electric method of inducing motion in liquid steel without using any mechanical device. In continuous casting EMS can be applied in mold and secondary cooling zone to

- Decrease segregation and porosity
- Improve steel cleanliness by forcing inclusions to float on the surface where they can be absorbed by a slag
- Improve the steel quality in terms of reducing the defects

Stirring induced by EMS modifies the flow pattern of molten steel of the solidifying strand. Improved stirring reduces segregation. Because of the imposed stirring, the superheat of the melt can be quickly dissipated which results in modification of structure of the solidifying strand from columnar to equiaxed and then to globular structure.

EMS allows to increase the casting speed which increases productivity of the caster.

An inductive electromagnetic stirrer is the stator of an asynchronomous motor, the rotor of which is the liquid core of the solidifying strand. This stator produces either a rotating or travelling magnetic field B, which induces eddy current J, perpendicular to B and its velocity vector. B and J combine together to create an electromagnetic force which causes stirring in the bath. The mains frequency is usually used for stirring billets or small blooms beneath the mold. In secondary cooling zone low frequency (1 to 20Hz0 is required to electromagnetically stir the liquid steel of the solidifying strand because the thick solid steel shell shields the magnetic flux.

High Speed Casting

High speed casting of slab increases the productivity. Average casting speed in the conventional slab casters is on average 2 m/min. Now if we want to increase the casting speed we have to consider design and operational features of tundish, mold and secondary cooling zone. This is illustrated by the following example:

Consider a conventional 2 strand caster casting slab of cross section 280mm x 1950mm at 2 m/min casting speed. Molten steel is fed from a tundish of 70 tons capacity by submerged entry nozzles in both the molds of the continuous casting machine.

Volume of slab /minute = 1.092 m^3/minute. Assuming density of liquid steel as 7 tons/m^3 steel flow rate is 7.6 tons/minute/strand.

Now if we increase casting speed to 3 m/min., the required steel flow rate would be 11.4 tons/min/strand, and which will increase to 15.2 tons/min/strand for 4 m/min casting speed. Now let us calculate the average residence time of molten steel flowing in the tundish assuming 70 tons tundish will be used even for high speed casting.

Steel flow rate (tons/min/strand)	Residence time (minutes)
7.6	4.6
11.4	3.1
15.2	2.3

We note that the average residence time of steel melt in the tundish decreases with the increase in the casting speed. Thus, chances of inclusion floatation will be very low at higher casting speeds. Thus, the following modifications may be considered;

In one modification tundish capacity needs to be increased, if tundish is to be used to remove inclusions during the process of continuous casting. Thus, a change in tundish design would be required. In other modification we improve the upstream steelmaking facility so that inclusion content in steel are within the tolerance limit. One may consider other alternatives depending on the available resources.

We have to consider further as to what will happen in the mold? Increase in steel flow rate will increase the steel velocity in the mould since the mold length is not being changed. One requires to consider the cross section of the submerged entry nozzle too. The higher steel flow rate in the mold would require intense cooling in the mold. The steel flow in the mold will be more turbulent. Also the mold powder consumption may increase. We would be requiring mold flux whose melting rate is relatively higher to keep pace with the casting speed.

Similarly water spray in the secondary cooling zone has to be modified in view of the increase in the casting speed. Uniform cooling of the strand is the prerequisite for the success of the caster.

Thin Slab Casting

Thin slab casting aims at to cast slabs of thicknesses less than 100 mm, say around 60 to 80 mm. The objective is to integrate caster, reheating furnace and hot strip mill to increase the productivity. A slab of cross section 1500mm x 80mm would require melt flow rate 3.78 tons/min at 4.5 m/min casting speed. Since the mould size is smaller turbulence in the mould increases which results in entrapment of mould flux on surface causing surface defects.

Use of funnel shaped mould with enlarged cross section at the meniscus is one remedy. Electro-magnetic brakes are very effective in parallel moulds.

Development of liquid core reduction is one such technology that has helped thin slab casting development. Solid ingot with liquid core is subjected to on-line rolling. This brings the following advantages:

- Rolling can be eliminated
- Improved internal quality in term of segregation

Thin slab caster can be integrated with EAF units and hence mini steel plants are using thin slab casting to produce hot strips with lower segregation, finer grain size and higher strength.

Near Net Shape Casting

The conventional methods of producing metallic strips and sheets require the casting of large ingots which are subsequently hot rolled and cold rolled to final thickness. Several unit operations like ingot soaking, slabbing mill, intermediate annealing, pickling are required to produce the strips. An alternate to this route is the near net shape casting or direct strip casting. In this technology strips are cast close to the final desired thickness and thus many of the unit operations in the conventional technology can be eliminated. The near net shape casting units are classified into four categories, namely

- Thin slab caster producing 20mm to 70mm thick slabs of cross section 20mm to 70mm X 1000mm to 2000mm which could be fed directly into the finishing stands of the hot strip mill without any conditioning.
- The thick strip caster producing 10mm to 20mm thick strips which may need some limited rolling for metallurgical reasons.
- Strip casters producing strips less than 10mm thickness which is sent directly to the cold strip mil.
- The thin strip or foil caster with a thickness of approximately 20μm to 500μm thickness and upto 300 mm wide.

Several technologies are developed for the above casters. In the technology developed by Mannesman demag Huettentechnik thin slab of 40mm to 70mm thickness and 1200 mm width can be produced. The features are:

- A narrow submerged nozzle
- Vertically curved mould with parallel broad faces that guide the strand vertically
- Adjustable mould width
- Dry casting technique

In addition there are twin roll caster, drum and ring caster, single roll caster, rheocaster and powder rolling caster.

Mist Spray Cooling

In mist spray, cooling is provided by a mist which is a two phase mixture of air and water; water droplets are finely distributed in the air jet. Air in which fine water droplets are distributed is blown on to the casting section; the water disperses on the casting section surface in the form of a mist. The mist spray provides uniform cooling. For the mist, control of volume flow rate of air to volume flow rate of water and upstream pressure must be suitably selected. The condition of the mist stream is strongly affected by the design of the nozzle.

Continuous Casting and Strip Casting

Molten steel is being cast continuously in to billets, blooms and slabs depending on the desired product i.e. whether long or flat products. In continuous casting, tundish, mold and secondary cooling sprays are arranged such that steel is poured continuously from the tundish and the solidified cast product is withdrawn continuously.

The arrangement of the tundish, mold and spray is shown in the figure.

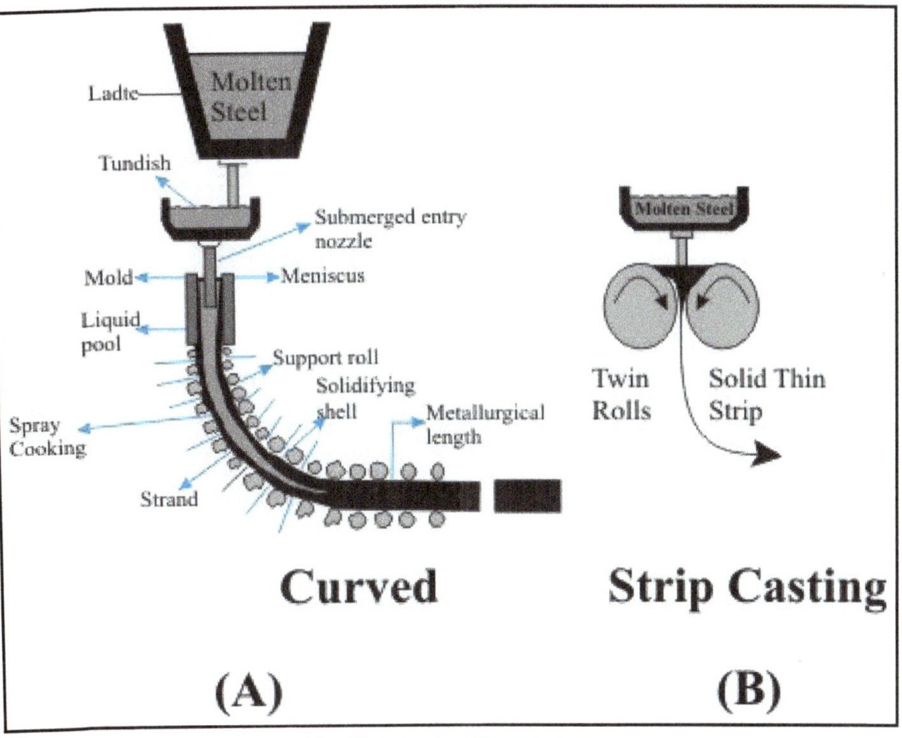

Continuous casting process

The original continuous casting machines were of vertical types. Now most of the continuous casters have either curved mould or vertical mold with bending rolls.

In the continuous casting, tundish is the important refractory lined vessel. It feeds the molten steel into the molds placed beneath the tundish through a submerged nozzle. Tundish also acts as reservoir of molten steel during ladle change-over periods and sequence casting. Modern tundishes are equipped with furniture like dams, weirs, slotted dams etc. to modify the molten steel flowing in the tundish during the process of continuous casting. Modern developments include thin slab caster, liquid core reduction. Thin slab casters are connected to the strip mill. The objective is to integrate the casting and rolling in order to save reheating cost.

Strip casting is also becoming popular in steel plants. Here molten steel is cast directly into the strip.

Final Finishing Operations

It has been considered appropriate to include final finishing operations in steelmaking course to appreciate integration between chemistry and cleanliness of steel and the final finishing operations. It is thought that the reader can appreciate the role of steelmaking in the product development and failure.

- Deformation processing technologies like forging rolling etc.
- Heat treatment to produce the finished product. Heat treatment consists of heating the steel products to a temperature in the austenitic region and then cooling.
- Surface hardening treatment.

References

- Degarmo, E. Paul; Black, J T.; Kohser, Ronald A. (2003), Materials and Processes in Manufacturing (9th ed.), Wiley, p. 277, ISBN 0-471-65653-4
- "'4th generation of thin slab casting technology - present and future' - MPT International". Stahleisen Communications. Retrieved 2014-03-29
- B. Ravi (1 January 2005). Metal Casting: Computer-Aided Design and Analysis. PHI Learning Pvt. Ltd. pp. 92–. ISBN 978-81-203-2726-9
- "'Modeling of Continuous Casting Defects Related to Mold Fluid Flow' - AIST Transactions, Iron and Steel Technology, July 2006" (PDF). AIST International. Retrieved 2014-03-29
- T F Waters (11 September 2002). Fundamentals of Manufacturing For Engineers. CRC Press. pp. 17–. ISBN 978-0-203-50018-7
- "'Transient Fluid-Flow Phenomena in the Continuous Steel-Slab Casting Mold and Defect Formation' - JOM Online". The Minerals, Metals, and Materials Society. Retrieved 2014-03-29
- Gordon Elliott (2006). Aspects of Ceramic History: A Series of Papers Focusing on the Ceramic Artifact As Evidence of Cultural and Technical Development. Gordon Elliott. pp. 52–. ISBN 978-0-9557690-0-9
- "'Analyzing Casting Problems by the On-line Monitoring of Continuous Casting Mold Temperatures' - JOM Online". The Minerals, Metals, and Materials Society. Retrieved 2014-03-29
- James T. Frane (1994). Craftsman's Illustrated Dictionary of Construction Terms. Craftsman Book Company. pp. 126–. ISBN 978-1-57218-008-6
- "Nucor Shatters Strip-Casting Record with the Castrip(R) Process'". Nucor Corporation, press release Jan 15, 2008. Retrieved 2014-03-29

6

Oxidation and Deoxidation Reactions in Steelmaking

During the process of making steel, there are many impurities present that need to be removed. Impurities such as carbon, silicon, phosphorus, manganese and sulphur are removed through oxidation. After the process, oxygen may still be left. This is removed through deoxidization. This chapter has been carefully written to provide an easy understanding of the varied facets of steelmaking reactions.

Redox

Sodium and fluorine bonding ionically to form sodium fluoride. Sodium loses its outer electron to give it a stable electron configuration, and this electron enters the fluorine atom exothermically. The oppositely charged ions are then attracted to each other. The sodium is oxidized, and the fluorine is reduced.

Redox (short for reduction–oxidation reaction) is a chemical reaction in which the oxidation states of atoms are changed. Any such reaction involves both a reduction process and a complementary oxidation process, two key concepts involved with electron transfer processes. Redox reactions include all chemical reactions in which atoms have their oxidation state changed; in general, redox reactions involve the transfer of electrons between chemical species. The chemical species from which the electron is stripped is said to have been oxidized, while the chemical species to which the electron is added is said to have been reduced. It can be explained in simple terms:

- Oxidation is the *loss* of electrons or an *increase* in oxidation state by a molecule, atom, or ion.
- Reduction is the *gain* of electrons or a *decrease* in oxidation state by a molecule, atom, or ion.

Reduction

Oxidant + e^- ⟶ Product

(Gain of Electrons) (Oxidation Number Decreases)

Oxidation

Reductant ⟶ Product + e^-

(Loss of Electrons) (Oxidation Number Increases)

The two parts of a redox reaction

Rust, a slow redox reaction

As an example, during the combustion of wood, oxygen from the air is reduced, gaining electrons from the carbon. Although oxidation reactions are commonly associated with the formation of oxides from oxygen molecules, oxygen is not necessarily included in such reactions, as other chemical species can serve the same function.

The reaction can occur relatively slowly, as in the case of rust, or more quickly, as in the case of fire. There are simple redox processes, such as the oxidation of carbon to yield carbon dioxide (CO_2) or the reduction of carbon by hydrogen to yield methane (CH_4), and more complex processes such as the oxidation of glucose ($C_6H_{12}O_6$) in the human body.

A bonfire; combustion is a fast redox reaction

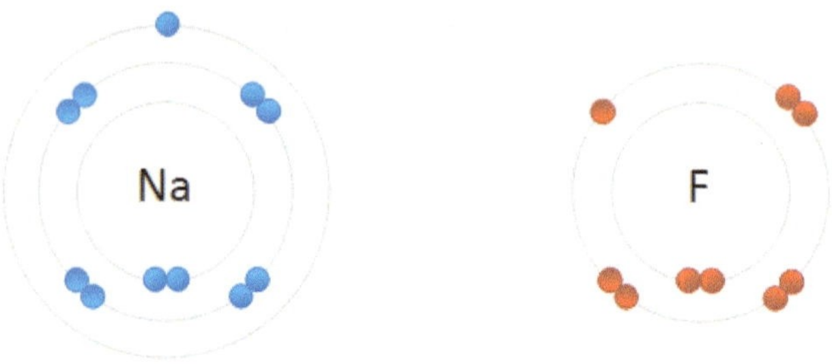

Sodium and fluorine bonding ionically to form sodium fluoride. Sodium loses its outer electron to give it a stable electron configuration, and this electron enters the fluorine atom exothermically. The oppositely charged ions are then attracted to each other. The sodium is oxidized, and the fluorine is reduced.

Etymology

"Redox" is a portmanteau of "reduction" and "oxidation".

The word *oxidation* originally implied reaction with oxygen to form an oxide, since dioxygen (O_2 (g)) was historically the first recognized oxidizing agent. Later, the term was expanded to encompass oxygen-like substances that accomplished parallel chemical reactions. Ultimately, the meaning was generalized to include all processes involving loss of electrons.

The word *reduction* originally referred to the loss in weight upon heating a metallic ore such as a metal oxide to extract the metal. In other words, ore was "reduced" to metal. Antoine Lavoisier (1743–1794) showed that this loss of weight was due to the loss of oxygen as a gas. Later, scientists realized that the metal atom gains electrons in this process. The meaning of *reduction* then became generalized to include all processes involving gain of electrons. Even though "reduction" seems counter-intuitive when speaking of the gain of electrons, it might help to think of reduction as the loss of oxygen, which was its historical meaning. Since electrons are negatively charged, it is also helpful to think of this as reduction in electrical charge.

The electrochemist John Bockris has used the words *electronation* and *deelectronation* to describe reduction and oxidation processes respectively when they occur at electrodes. These words are analogous to protonation and deprotonation, but they have not been widely adopted by chemists.

The term "hydrogenation" could be used instead of reduction, since hydrogen is the reducing agent in a large number of reactions, especially in organic chemistry and biochemistry. But, unlike oxidation, which has been generalized beyond its root element, hydrogenation has maintained its specific connection to reactions that *add* hydrogen to another substance (e.g., the hydrogenation of unsaturated fats into saturated fats, $R{-}CH{=}CH{-}R + H_2 \rightarrow R{-}CH_2{-}CH_2{-}R$). The word "redox" was first used in 1928.

Definitions

The processes of oxidation and reduction occur simultaneously and cannot happen independently of one another, similar to the acid–base reaction. The oxidation alone and the reduction alone are each called a *half-reaction*, because two half-reactions always occur together to form a whole reaction. When writing half-reactions, the gained or lost electrons are typically included explicitly in order that the half-reaction be balanced with respect to electric charge.

Though sufficient for many purposes, these general descriptions are not precisely correct. Although oxidation and reduction properly refer to *a change in oxidation state* — the actual transfer of electrons may never occur. The oxidation state of an atom is the fictitious charge that an atom would have if all bonds between atoms of different elements were 100% ionic. Thus, oxidation is best defined as an *increase in oxidation state*, and reduction as a *decrease in oxidation state*. In practice, the transfer of electrons will always cause a change in oxidation state, but there are many reactions that are classed as "redox" even though no electron transfer occurs (such as those involving covalent bonds).

Oxidizing and Reducing Agents

In redox processes, the reductant transfers electrons to the oxidant. Thus, in the reaction, the reductant or *reducing agent* loses electrons and is oxidized, and the oxidant or *oxidizing agent* gains electrons and is reduced. The pair of an oxidizing and reducing agent that are involved in a particular reaction is called a redox pair. A redox couple is a reducing species and its corresponding oxidizing form, e.g., Fe^{2+}/Fe^{3+}.

Oxidizers

The international pictogram for oxidizing chemicals.

Substances that have the ability to oxidize other substances (cause them to lose electrons) are said to be oxidative or oxidizing and are known as oxidizing agents, oxidants, or oxidizers. That is, the oxidant (oxidizing agent) removes electrons from another substance, and is thus itself reduced. And, because it "accepts" electrons, the oxidizing agent is also called an electron acceptor. Oxygen is the quintessential oxidizer.

Oxidants are usually chemical substances with elements in high oxidation states (e.g., H_2O_2, MnO_4^-, CrO_3, $Cr_2O_7^{2-}$, OsO_4), or else highly electronegative elements (O_2, F_2, Cl_2, Br_2) that can gain extra electrons by oxidizing another substance.

Reducers

Substances that have the ability to reduce other substances (cause them to gain electrons) are said to be reductive or reducing and are known as reducing agents, reductants, or reducers. The reductant (reducing agent) transfers electrons to another substance, and is thus itself oxidized. And, because it "donates" electrons, the reducing agent is also called an electron donor. Electron donors can also form charge transfer complexes with electron acceptors.

Reductants in chemistry are very diverse. Electropositive elemental metals, such as lithium, sodium, magnesium, iron, zinc, and aluminium, are good reducing agents. These metals donate or *give away* electrons readily. *Hydride transfer reagents*, such as $NaBH_4$ and $LiAlH_4$, are widely used in organic chemistry, primarily in the reduction of carbonyl compounds to alcohols. Another method of reduction involves the use of hydrogen gas (H_2) with a palladium, platinum, or nickel catalyst.

These *catalytic reductions* are used primarily in the reduction of carbon-carbon double or triple bonds.

Standard Electrode Potentials (Reduction Potentials)

Each half-reaction has a *standard electrode potential* (E^o_{cell}), which is equal to the potential difference or voltage at equilibrium under standard conditions of an electrochemical cell in which the cathode reaction is the half-reaction considered, and the anode is a standard hydrogen electrode where hydrogen is oxidized:

$$\frac{1}{2}H_2 \rightarrow H^+ + e^-.$$

The electrode potential of each half-reaction is also known as its *reduction potential* E^o_{red}, or potential when the half-reaction takes place at a cathode. The reduction potential is a measure of the tendency of the oxidizing agent to be reduced. Its value is zero for $H^+ + e^- \rightarrow \frac{1}{2} H_2$ by definition, positive for oxidizing agents stronger than H^+ (e.g., +2.866 V for F_2) and negative for oxidizing agents that are weaker than H^+ (e.g., −0.763 V for Zn^{2+}).

For a redox reaction that takes place in a cell, the potential difference is:

$$E^0_{cell} = E^0_{cathode} - E^0_{anode}$$

However, the potential of the reaction at the anode was sometimes expressed as an *oxidation potential*:

$$E^0_{ox} = -E^0_{red}$$

The oxidation potential is a measure of the tendency of the reducing agent to be oxidized, but does not represent the physical potential at an electrode. With this notation, the cell voltage equation is written with a plus sign

$$E^0_{cell} = E^0_{red(cathode)} + E^0_{ox(anode)}$$

Examples of Redox Reactions

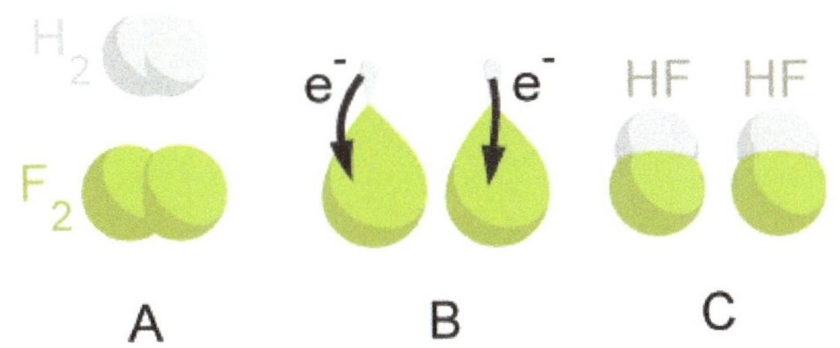

Illustration of a redox reaction

A good example is the reaction between hydrogen and fluorine in which hydrogen is being oxidized and fluorine is being reduced:

$$H_2 + F_2 \rightarrow 2HF$$

We can write this overall reaction as two half-reactions:

the oxidation reaction:

$$H_2 + 2H^+ \rightarrow 2\,e^-$$

and the reduction reaction:

$$F_2 + 2e^- \rightarrow 2F^-$$

Analyzing each half-reaction in isolation can often make the overall chemical process clearer. Because there is no net change in charge during a redox reaction, the number of electrons in excess in the oxidation reaction must equal to the number consumed by the reduction reaction.

Elements, even in molecular form, always have an oxidation state of zero. In the first half-reaction, hydrogen is oxidized from an oxidation state of zero to an oxidation state of +1. In the second half-reaction, fluorine is reduced from an oxidation state of zero to an oxidation state of −1.

When adding the reactions together the electrons are canceled:

$$H_2 \rightarrow 2H^+ + 2e^-$$

$$\underline{F_2 + 2e^- \rightarrow 2F^-}$$

$$H_2 + F_2 \rightarrow 2H^+ + 2F^-$$

And the ions combine to form hydrogen fluoride:

$$2H^+ + 2F^- \rightarrow 2HF$$

The overall reaction is:

$$H_2 + F_2 \rightarrow 2HF$$

Metal Displacement

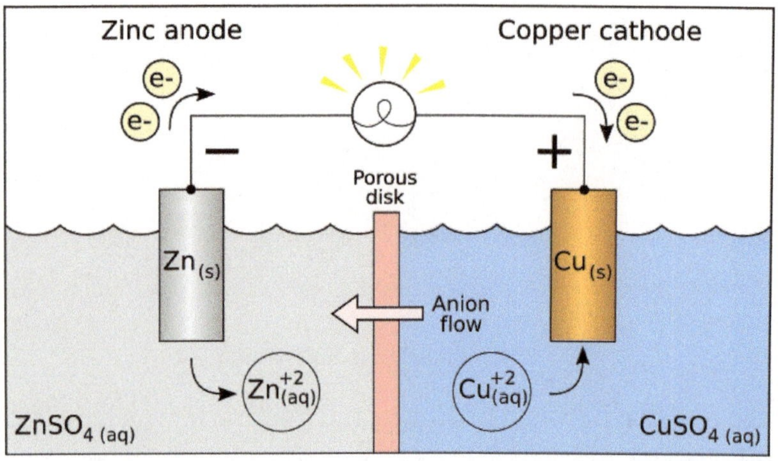

A redox reaction is the force behind an electrochemical cell like the Galvanic cell pictured. The battery is made out of a zinc electrode in a $ZnSO_4$ solution connected with a wire and a porous disk to a copper electrode in a $CuSO_4$ solution.

In this type of reaction, a metal atom in a compound (or in a solution) is replaced by an atom of another metal. For example, copper is deposited when zinc metal is placed in a copper(II) sulfate solution:

$$Zn(s) + CuSO_4(aq) \rightarrow ZnSO_4(aq) + Cu(s)$$

In the above reaction, zinc metal displaces the copper(II) ion from copper sulfate solution and thus liberates free copper metal.

The ionic equation for this reaction is:

$$Zn + Cu^{2+} \rightarrow Zn^{2+} + Cu$$

As two half-reactions, it is seen that the zinc is oxidized:

$$Zn \rightarrow Zn^{2+} + 2e^-$$

And the copper is reduced:

$$Cu^{2+} + 2e^- \rightarrow Cu$$

Other examples

- The reduction of nitrate to nitrogen in the presence of an acid (denitrification):

$$2NO_3^- + 10e^- + 12H^+ \rightarrow N_2 + 6H_2O$$

- The combustion of hydrocarbons, such as in an internal combustion engine, which produces water, carbon dioxide, some partially oxidized forms such as carbon monoxide, and heat energy. Complete oxidation of materials containing carbon produces carbon dioxide.
- In organic chemistry, the stepwise oxidation of a hydrocarbon by oxygen produces water and, successively, an alcohol, an aldehyde or a ketone, a carboxylic acid, and then a peroxide.

Corrosion and rusting

Oxides, such as iron(III) oxide or rust, which consists of hydrated iron(III) oxides $Fe_2O_3 \cdot nH_2O$ and iron(III) oxide-hydroxide ($FeO(OH)$, $Fe(OH)_3$), form when oxygen combines with other elements

Iron rusting in pyrite cubes

- The term corrosion refers to the electrochemical oxidation of metals in reaction with an oxidant such as oxygen. Rusting, the formation of iron oxides, is a well-known example of electrochemical corrosion; it forms as a result of the oxidation of iron metal. Common rust often refers to iron(III) oxide, formed in the following chemical reaction:

$$4Fe + 3O_2 \rightarrow 2Fe_2O_3$$

- The oxidation of iron(II) to iron(III) by hydrogen peroxide in the presence of an acid:

$$Fe^{2+} \rightarrow Fe^{3+} + e^-$$
$$H_2O_2 + 2e^- \rightarrow 2OH^-$$

Overall equation:

$$2Fe^{2+} + H_2O_2 + 2H^+ \rightarrow 2Fe^{3+} + 2H_2O$$

Redox Reactions in Industry

Cathodic protection is a technique used to control the corrosion of a metal surface by making it the cathode of an electrochemical cell. A simple method of protection connects protected metal to a more easily corroded "sacrificial anode" to act as the anode. The sacrificial metal instead of the protected metal, then, corrodes. A common application of cathodic protection is in galvanized steel, in which a sacrificial coating of zinc on steel parts protects them from rust.

The primary process of reducing ore at high temperature to produce metals is known as smelting.

Oxidation is used in a wide variety of industries such as in the production of cleaning products and oxidizing ammonia to produce nitric acid, which is used in most fertilizers.

Redox reactions are the foundation of electrochemical cells, which can generate electrical energy or support electrosynthesis.

The process of electroplating uses redox reactions to coat objects with a thin layer of a material, as in chrome-plated automotive parts, silver plating cutlery, and gold-plated jewelry.

The production of compact discs depends on a redox reaction, which coats the disc with a thin layer of metal film.

Redox Reactions in Biology

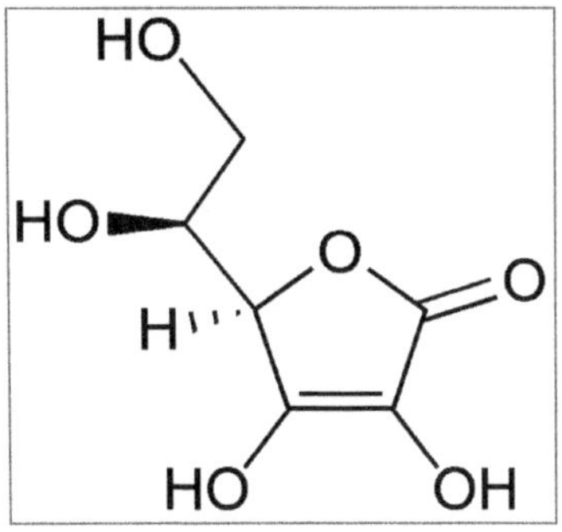

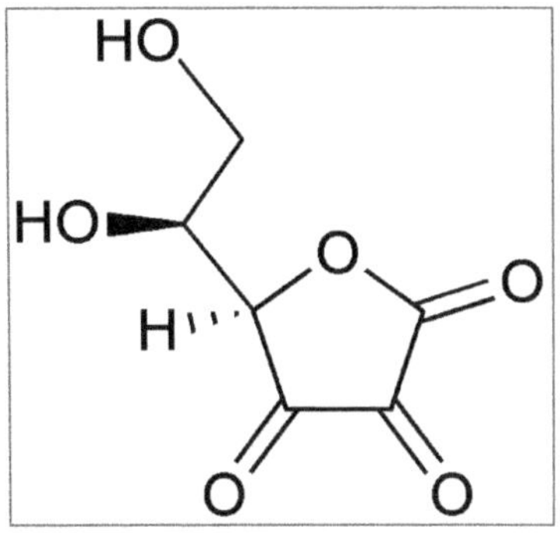

Top: ascorbic acid (reduced form of Vitamin C)
Bottom: dehydroascorbic acid (oxidized form of Vitamin C)

Many important biological processes involve redox reactions.

Cellular respiration, for instance, is the oxidation of glucose ($C_6H_{12}O_6$) to CO_2 and the reduction of oxygen to water. The summary equation for cell respiration is:

$$C_6H_{12}O_6 + 6\ O_2 \rightarrow 6\ CO_2 + 6\ H_2O$$

The process of cell respiration also depends heavily on the reduction of NAD^+ to NADH and the reverse reaction (the oxidation of NADH to NAD^+). Photosynthesis and cellular respiration are complementary, but photosynthesis is not the reverse of the redox reaction in cell respiration:

$$6\ CO_2 + 6\ H_2O + \text{light energy} \rightarrow C_6H_{12}O_6 + 6\ O_2$$

Biological energy is frequently stored and released by means of redox reactions. Photosynthesis involves the reduction of carbon dioxide into sugars and the oxidation of water into molecular oxygen. The reverse reaction, respiration, oxidizes sugars to produce carbon dioxide and water. As intermediate steps, the reduced carbon compounds are used to reduce nicotinamide adenine dinucleotide (NAD^+), which then contributes to the creation of a proton gradient, which drives the synthesis of adenosine triphosphate (ATP) and is maintained by the reduction of oxygen. In animal cells, mitochondria perform similar functions.

Free radical reactions are redox reactions that occur as a part of homeostasis and killing microorganisms, where an electron detaches from a molecule and then reattaches almost instantaneously. Free radicals are a part of redox molecules and can become harmful to the human body if they do not reattach to the redox molecule or an antioxidant. Unsatisfied free radicals can spur the mutation of cells they encounter and are, thus, causes of cancer.

The term redox state is often used to describe the balance of GSH/GSSG, NAD^+/NADH and $NADP^+$/NADPH in a biological system such as a cell or organ. The redox state is reflected in the balance of several sets of metabolites (e.g., lactate and pyruvate, beta-hydroxybutyrate, and acetoacetate), whose interconversion is dependent on these ratios. An abnormal redox state can develop in a variety of deleterious situations, such as hypoxia, shock, and sepsis. Redox mechanism also control some cellular processes. Redox proteins and their genes must be co-located for redox regulation according to the CoRR hypothesis for the function of DNA in mitochondria and chloroplasts.

Redox Cycling

A wide variety of aromatic compounds are enzymatically reduced to form free radicals that contain one more electron than their parent compounds. In general, the electron donor is any of a wide variety of flavoenzymes and their coenzymes. Once formed, these anion free radicals reduce molecular oxygen to superoxide, and regenerate the unchanged parent compound. The net reaction is the oxidation of the flavoenzyme's coenzymes and the reduction of molecular oxygen to form superoxide. This catalytic behavior has been described as futile cycle or redox cycling.

Examples of redox cycling-inducing molecules are the herbicide paraquat and other viologens and quinones such as menadione.

Redox Reactions in Geology

Mi Vida uranium mine, near Moab, Utah. The alternating red and white/green bands of sandstone correspond to oxidized and reduced conditions in groundwater redox chemistry.

In geology, redox is important to both the formation of minerals and the mobilization of minerals, and is also important in some depositional environments. In general, the redox state of most rocks can be seen in the color of the rock. The rock forms in oxidizing conditions, giving it a red color. It is then "bleached" to a green—or sometimes white—form when a reducing fluid passes through the rock. The reduced fluid can also carry uranium-bearing minerals. Famous examples of redox conditions affecting geological processes include uranium deposits and Moqui marbles.

Balancing Redox Reactions

Describing the overall electrochemical reaction for a redox process requires a balancing of the component half-reactions for oxidation and reduction. In general, for reactions in aqueous solution, this involves adding H^+, OH^-, H_2O and electrons to compensate for the oxidation changes.

Acidic Media

In acidic media, H^+ ions and water are added to half-reactions to balance the overall reaction.

For instance, when manganese(II) reacts with sodium bismuthate:

$$\text{Unbalanced reaction}: Mn^{2+}(aq) + NaBiO_3(s) \rightarrow Bi^{3+}(aq) + MnO_4^-(aq)$$

$$\text{Oxidation}: \quad 4H_2O(l) + Mn^{2+}(aq) \rightarrow MnO_4^-(aq) + 8H^+(aq) + 5e^-$$

$$\text{Reduction}: \quad 2e^- + 6H^+ + BiO_3^-(s) \rightarrow Bi^{3+}(aq) + 3H_2O(l)$$

The reaction is balanced by scaling the two half-cell reactions to involve the same number of electrons (multiplying the oxidation reaction by the number of electrons in the reduction step and vice versa):

$$8H_2O(l) + 2Mn^{2+}(aq) \rightarrow 2MnO_4^-(aq) + 16H^+(aq) + 10e^-$$

$$10e^- + 30H^+ + 5BiO_3^-(s) \rightarrow 5Bi^{3+}(aq) + 15H_2O(l)$$

Adding these two reactions eliminates the electrons terms and yields the balanced reaction:

$$14H^+(aq) + 2Mn^{2+}(aq) + 5NaBiO_3(s) \rightarrow 7H_2O(l) + 2MnO_4^-(aq) + 5Bi^{3+}(aq) + 5Na^+(aq)$$

Basic Media

In basic media, OH^- ions and water are added to half reactions to balance the overall reaction.

For example, in the reaction between potassium permanganate and sodium sulfite:

$$\text{Unbalanced reaction}: \quad KMnO_4 + Na_2SO_3 + H_2O \rightarrow MnO_2 + Na_2SO_4 + KOH$$

$$\text{Reduction}: \quad 3e^- + 2H_2O + MnO_4^- \rightarrow MnO_2 + 4OH^-$$

$$\text{Oxidation}: \quad 2OH^- + SO_3^{2-} \rightarrow SO_4^{2-} + H_2O + 2e^-$$

Balancing the number of electrons in the two half-cell reactions gives:

$$6e^- + 4H_2O + 2MnO_4^- \rightarrow 2MnO_2 + 8OH^-$$

$$6OH^- + 3SO_3^{2-} \rightarrow 3SO_4^{2-} + 3H_2O + 6e^-$$

Adding these two half-cell reactions together gives the balanced equation:

$$2KMnO_4 + 3Na_2SO_3 + H_2O \rightarrow 2MnO_2 + 3Na_2SO_4 + 2KOH$$

Memory Aids

The key terms involved in redox are often confusing to students. For example, an element that is oxidized loses electrons; however, that element is referred to as the reducing agent. Likewise, an element that is reduced gains electrons and is referred to as the oxidizing agent. Acronyms or mnemonics are commonly used to help remember the terminology:

- "OIL RIG" — oxidation is loss of electrons, reduction is gain of electrons.
- "LEO the lion says GER" — loss of electrons is oxidation, gain of electrons is reduction.
- "LEORA says GEROA" — loss of electrons is oxidation (reducing agent), gain of electrons is reduction (oxidizing agent).
- "RED CAT" and "AN OX", or "AnOx RedCat" ("an ox-red cat") — reduction occurs at the cathode and the anode is for oxidation.
- "RED CAT gains what AN OX loses" – reduction at the cathode gains (electrons) what anode oxidation loses (electrons).

Oxidation Reactions

The principle reactions in steelmaking comprise of oxidation of impurity elements by oxygen dissolved in hot metal or FeO content of slag.

$$[Fe]+[O]=(FeO)$$
$$[Mn]+[O]=(MnO)$$
$$[Si]+[O]=(SiO_2)$$
$$[C]+[O]=\{CO\}$$
$$2[P]+5[O]=(P_2O_5)$$

- All reactions are exothermic.
- C is removed as gas.
- Except C, all other impurities are removed as oxides and all these oxides float on the surface of the molten metal during refining of hot metal to steel.
- Iron oxidation is unavoidable. Oxidation of Fe is loss in productivity; hence its oxidation must be controlled.
- Oxygen must be dissolved to remove an impurity from the hot metal.

We begin with considering oxidation of an individual element and evolve the optimum conditions using thermodynamic principles.

Iron Oxidation

Oxidation of iron i.e. the most important since it controls:

- FeO content of slag and oxygen content of steel
- Loss of iron in slag and hence affects productivity
- Oxidation potential of slag

In addition to the above FeO also helps in dissolution of lime in slag.

Consider the reaction

$$[Fe]+[O]= (FeO)$$

The equilibrium constant K_{Fe} is

$$K_{Fe} = \frac{a_{FeO}}{a_{FehO}} = \frac{a_{(FeO)}}{a_{[Fe]} f_O [wt\%O]}$$

h is henrian and a is raoultian activity. Since Fe in steel is pure; $a_{Fe} = 1$, and

$$\log f_O = -0.17[wt\%O]$$

$$\log K_{Fe} = \frac{6150}{T} - 2.604$$

In equation above, T is in K. By equations above we get

$$[wt\%O]10^{-0.17[wt\%O]} = [a_{Feo}\{10^{(\frac{6150}{T}-2.604)}\}^{-1}]$$

The equation above can be used to determine wt% O in steel at any temperature T, when a_{FeO} in slag is known.

When pure FeO is in contact with $a_{FeO} = 1$. We can determine wt%O at saturation for different temperatures:

T (K)	[wt%O]cat.
1873	0.233
1923	0.285

We note that increase in temperature increases oxygen dissolved in molten iron.

The above values of dissolved oxygen correspond when pure FeO is in contact with Fe pure. In steelmaking FeO is present along with other oxides like calcium oxide, magnesium oxide, silica, manganese oxide etc, hence activity of FeO is influenced by other solute oxides. Thus $a_{FeO} = \gamma_{FeO} N_{FeO}$

where γ_{FeO} is activity coefficient and N_{FeO} mole fraction of FeO in slag, γ_{FeO} depends on slag composition. In CaO-SiO_2-FeO system, as CaO/SiO_2 ratio increases a_{FeO} increases; physically it means that CaO replaces FeO from FeO. SiO_2. The following expression is used to express a_{FeO}:

$$a_{FeO} = 0.514(N_{FeO})^{0.2665}$$

Consider a slag with $N_{FeO} = 0.5$; a_{FeO} according to equation 10 is 0.31; [wt% O] in steel would be 0.072 as calculated by equation above.

Few other equations are available; i.e.

$$\log[\%O]_{Sat.} = -\frac{6320}{T} + 2.734$$

$$\log[\%O]_{Sat.} = -\frac{6400}{T} + 2.756$$

The calculations are made on [wt% O] by equations above at different temperatures using $a_{FeO} = 1$.

T (K)	$[wt\%\ O]_9$	$[wt\%\ O]_{11}$	$[wt\%\ O]_{12}$
1873	0.233	0.229	0.220
1923	0.285	0.280	0.268

There is a slight difference in the values of dissolved oxygen content in steel. But all equations suggest that increase in temperature increases dissolved oxygen in iron which is in contact with pure FeO. This calculation indicates that control of temperature is important to limit the dissolution of oxygen in molten iron.

Oxidation of Silicon

Consider reaction above

$$[Si] + 2[O] = (SiO_2)$$

$$K_{Si} = \frac{a_{(SiO_2)}}{[wt\%Si][wt\%O]^2}$$

$$[wt\%Si] \times [wt\%O]^2 = \frac{a_{(SiO_2)}}{K_{Si}}$$

Different sources give the following expression for K_{Si}

$$\log K_{Si} = \frac{30110}{T} - 11.4$$

$$\log K_{Si} = \frac{29700}{T} - 11.25$$

Both equations predict that decrease in temperature increases K_{Si}. There is a slight difference in values of K_{Si}. Equation above predicts 15-17% higher K_{Si}.

Conditions Favourable for Silicon Oxidation are

- Low temperature

- Low a_{SiO_2} in slag. A basic slag favours silicon oxidation.

In a basic slag, silicon oxidation occurs practically to a very low value since SiO_2 reacts with CaO and decreases activity of silica in slag.

Another important feature of silicon reaction is very high affinity of silicon with oxygen, silicon can be used as a deoxidizing agent.

By equation earlier

$$[wt\%O] = \sqrt{\frac{a_{SiO_2}}{K_{Si} \times [wt\%Si]}}$$

At 1773 K, $K_{Si} = 3.24 \times 10^5$ by equation earlier and using $a_{SiO2} = 1$, we get

$$[wt\%] = \sqrt{\frac{0.308 \times 10^{-5}}{[wt\%Si]}}$$

Equation shows drastic reduction in oxygen content of steel due to addition of silicon. This suggests that silicon is a very effective deoxidizer. Normally silicon is used as ferrosilicon in steelmaking.

Behavoiur of manganese in iron-carbon melt:

- Mn is soluble in iron in any proportion
- Mn forms ideal solutions in iron
- Carbon lowers the activity of Mn in Fe-Mn-C system by forming Mn_3C .

Oxidation of Manganese

Mn is oxidized readily at relatively low temperatures and can form oxides like MnO, MnO_2, Mn_2O_3 etc. But MnO is stable at high temperature.

$$[Mn] + O = (MnO)$$
$$[Mn] + (FeO) = (MnO) + [Fe]$$

The reaction 1 occurs with dissolved oxygen in metal, whereas reaction 2 is a slag/metal reaction. Both reactions are exothermic. Lower temperature favours oxidation of Mn from metal to slag; whereas higher temperature favours reduction of MnO of slag and there occurs reversal of Mn into steel. Reduction of MnO in slag is important; we consider reaction earlier

$$K = \frac{a_{(MnO)}\, a_{[Fe]}}{h_{[Mn]}\, a_{(FeO)}}$$

Replacing activity by mole fraction and using, $a_{[Fe]} = 1$ we get,

$$K \quad \frac{(MnO)\ (MnO)}{Mn[wt\%Mn] \times \gamma(FeO)\ (FeO)}$$

Grouping all activity coefficient terms and putting $N_{(MnO)} \approx (wt\%Mn)$ We get,

$$K^* = \frac{K(\gamma FeO)fMn}{\gamma MnO} = \frac{(wt\%Mn)}{[wt\%Mn] \times N_{FeO}}$$

Where K^* is an equilibrium quotient and it depends on composition of slag. Distribution of Mn between slag and metal can be written as

$$\varphi = \frac{(wt\%Mn)}{[wt\%Mn]} = K^* N_{FeO}$$

$$\log K^* = \frac{7940}{T} - 3.17$$

According to equation 7 K^* increases with decrease in temperature (K^* = 9.1, 11.72 and 20.33 at temperatures 1923K, 1873K and 1773K respectively).

Condition for Oxidation of Mn According to Equation above

- High activity of FeO in slag which means an oxidizing slag
- Decrease in temperature increases K* according to equation above.

Reduction of Mn in Slag

Conditions for reduction of MnO, that is reversal of reaction 2 is important. The reduction of MnO in slag transfers Mn from slag to metal and increases the concentration of manganese. Following are the conditions for the reduction of MnO in slag:

- Low activity of FeO in slag which means a reducing slag
- High temperature which decreases K^*

Illustration:

Consider a slag of basicity 1.8. At this basicity the activity coefficient of MnO in slag is 1.6. The mole fraction of FeO and MnO in slag is 0.25 and 0.05 respectively. Determine the equilibrium content of Mn and O in steel at 1873K. Given

$$[Mn] + (FeO) = (MnO) + [Fe]; \Delta G^\circ = -27800 + 11.8T$$

Using equilibrium constant definition, we can write

$$\text{In}[\%Mn] = \frac{\Delta G^\circ}{RT} + \text{In}\frac{a_{Fe\gamma MnO} N_{MnO}}{a_{FeO}}$$

Substituting the values, we get at 1873K

$$[\%Mn] = 0.048\%$$

using equation

$$a_{Feo} = 0.514(N_{Feo})^{0.2665}$$

We get $a_{Feo} = 0.36$

$$\text{Hence}\,[wt\%O] = [wt\%O]_{sat} \times a_{Feo}$$

$$= 0.233 \times 0.36$$
$$= 0.084\%$$

Calculations performed at 1773K shows that [wt% Mn] is 0.032. This means that decrease in temperature favours removal of manganese from metal to slag. The reader may perform calculations at 1973K and interpret the calculations.

Oxidation of Carbon

It is important to note that amongst all steelmaking reactions, oxidation of carbon is the reaction whose product is gas i.e. CO. Therefore this reaction is of very much significance during steelmaking because

- CO gas during escape from the molten bath can induce stirring in metal and slag phases during steelmaking.
- CO gas can cause slag to foam which leads to increase in surface area.
- CO gas has a high calorific value and combustion of CO in steelmaking can contribute to energy efficiency.

Carbon oxidation is also known as "decarburizing" reaction

$$[C] + [O] = \{CO\}$$

$$K_{CO} = \frac{p_{CO}}{h_C h_O} = \frac{p_{CO}}{[wt\%C] \times f_C \times [wt\%O] \times f_O}$$

$$[wt\%C] \times [wt\%O] = \frac{p_{CO}}{K_{CO}} \times \frac{1}{f_C \times f_O}$$

If we assume $f_C = f_O$ = unity that is at low concentration of carbon and oxygen in molten metal then

$$[wt\%C] \times [wt\%O] = \frac{p_{CO}}{K_{CO}}$$

According to eq. above, the product $[wt\%C] \times [wt\%O]$ at a given temperature depends only on partial pressure of CO in equilibrium with melt. It is important to note that p_{CO} depends on the location of nucleation of CO in steel melt. If CO nucleates deep into the bath then p_{CO} will be greater than atmospheric pressure.

Let us calculate equilibrium content of carbon and oxygen at 1873K for $p_{CO} = 1$ atm 1.2 atm and 1.5 atm

The value of K_{CO} is calculated from

$$\log K_{CO} = \frac{1056}{T} + 2.13$$

[wt% C]	[wt% O]		
	Pco = 1	Pco = 1.2	Pco = 1.5
0.05	0.0405	0.0486	0.0608
0.1	0.0202	0.0242	0.0303
0.5	0.0040	0.0048	0.0060
1.0	0.0020	0.0024	0.0030

From the Table we Note that

- Decrease in carbon content increases the oxygen dissolved in steel. This is important in connection with production or ultra low carbon steel for certain applications. Production of ultra low carbon steels will be accompanied with dissolved oxygen if precautions are not taken during steelmaking.
- Increase in P_{CO} increases [wt% O] in steel

Let us consider the evolution of CO gas. According to equation earlier,

12 Kg C produces 22.4 m^3 CO (1 atm and 273K)

1 Kg C produces 1.87 m^3 CO (1 atm and 273K) which is equivalent to 12.83 m^3 CO (1 atm and 1873 K)

Now for 1000 Kg Hot metal and 0.2% Carbon in Steel

CO production would be 488 m^3 (1atm, 1873 K) / ton of hot metal. This volume of CO will evolve no doubt over a period of time but at any time large amount of CO will be escaping the system. Escaping of this gas will agitate the bath and contribute to enhanced rates of mass transfer reactions. Also care must be taken for the easy and unhindered escape of CO gas from the vessel failing which foaming and eventually expulsion of slag may occur.

Rimming Reaction

Other aspect of carbon reaction is the evolution of CO during solidification of steel. As the temperature of molten steel decreases from 1873K to 1773K, K_{CO} increases from 494 to 532 which results in decrease in $[wt\%C] \times [wt\%O]$ as steel cools. This will lead to CO evolution during solidification and is called rimming reaction. Rimming reaction induces stirring in the solidifying liquid steel and minimizes segregation of solutes.

Dephosphorization Reaction

Equilibrium Considerations:

Phosphorus removal reaction

$$2[P]+5[O]=(P_2O_5)$$

$$\Delta G^\circ = -740375+535.365T\ \text{J/mol}$$

At T > 1382K, ΔG° becomes positive which results in decomposition of (P_2O_5) to P and O. Thus removal of phosphorus requires that $a_{P_2O_5}$ must be reduced.

$$Kp=\frac{a_{P_2O_5}}{[\text{wt\%P}]^2[\text{wt\%O}]^5}$$

Now

$$[\text{wt\%O}]=a_{FeO}[\text{wt\%O}]_{sat}\ \text{and}$$

$$\log[\text{wt\%O}]_{sat}=-\frac{6320}{T}+2.734$$

By equation above and replacing $a_{P_2O_5}$ by using Raoult's Law and after rearrangement

$$\frac{N_{(P_2O_5)}}{[\text{wt\%P}]^2}=\frac{K_p(a_{FeO})^5[\text{wt\%O}]_{sat}}{\gamma P_2O_5}$$

γP_2O_5 is activity coefficient of P_2O_5 in slag. The LHS of equation above is index of dephosphorization and denotes distribution of phosphorus between slag and metal. Higher value of LHS demands low γP_2O_5 in a slag of a given composition.

How low γP_2O_5 should be?

Consider dephosphorization in a slag of $a_{FeO}=0.31$ at 1773K. Initial %P in metal is 0.1 and mole fraction of $N_{P_2O_5}$ in slag = 0.01. Let us calculate $\gamma_{P_2O_5}$ which will allow dephosphorization.

$$\log K_p=\frac{38668}{T}-27.96$$

At 1773 K, $K_p=7.06\times10^{-7}$

$[\text{wt\%O}]$ can be determined by equation earlier. We substitute the values in equation above. We get $\gamma_{P_2O_5}=4.16\times10^{-16}$. Now the question before us is how to attain such a low value of $\gamma_{P_2O_5}$ in a slag of given composition? Such a low value of $\gamma_{P_2O_5}$ can be attained when we use basic oxides which have a very strong tendency to form a stable chemical compound. The different basic oxides have different ability to lower $\gamma_{P_2O_5}$. The following expression describes the relative effects of basic oxides on .

$$\log\gamma_{P_2O_5} = -24.64\left[N_{cao} + 0.682N_{mgo} + 0.591N_{mno} + 0.545_{Feo} - 0.091N_{SiO_2}\right] - \frac{42000}{T} + 23.58$$

Alkaline oxides Na_2O and BaO are stronger than CaO but they are corrosive to the refractory lining and hence not used.

Consider a slag $N_{CaO} = 0.56$, $N_{MgO} = 0.12$, $N_{MnO} = 0.06$, $N_{FeO} = 0.1$ and $N_{SiO_2} = 0.6$

We calculate $\gamma_{P_2O_5}$ at different temperatures

T (K)	$\gamma_{P_2O_5}$
1773	1.74×10^{-18}
1823	0.778×10^{-18}
1873	0.324×10^{-18}

Decrease in temperature increases $\gamma_{P_2O_5}$ which favours dephosphorization reaction.

Effect of FeO and CaO on Dephosphorization

Figure: shows the variation of dephosphorization index $(wt\%P_2O_5)/[wt\%P]$ as a function of wt%FeO for CaO-FeO- SiO_2 slag at different basicities. The dephosphorization ratio increases with

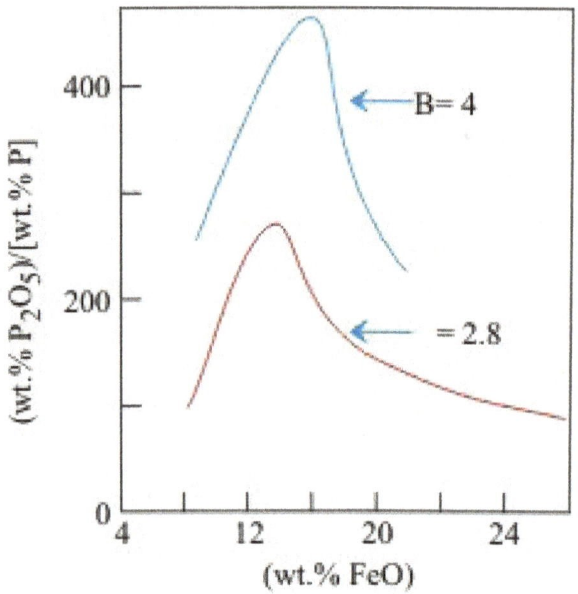

Dephosphorization index as a function of weight percent FeO content of slag for different basicities.

increase in FeO content of slag and becomes maximum in between 15-16% FeO at all basicities. Further increase in FeO beyond 15-16%, dephosphorization decreases. The above behaviour can be observed at all basicities of slag.

The above behaviour is due to the dual role of FeO. FeO is the source of oxygen for oxidation of P according to the following reaction

$$2[P]+(FeO)=(P_2O_5)+[Fe]$$

For a given basicity of slag, as FeO content of slag increases oxidizing power of slag increases and phosphorus oxidation according to reaction 8 will be favoured because CaO of slag decreases the activity of P_2O_5 by forming a stable compound. Beyond the optimum value of FeO in slag FeO replaces CaO and may either combine with CaO or with P_2O_5 . FeO is a weak base compared with CaO as a result of which the dephosphorization ratio decreases with addition of FeO beyond an optimum value.

The maximum dephosphorization ratio increases with the increase in the basicity of slag as can be seen in the figure. Higher basicity requires higher amount of CaO dissolved in slag. Any undissolved CaO will not be effective for dephosphorization. Optimum value of FeO is more or less independent of the basicity of slag. Thus control of FeO in slag is important for efficient dephhosphorization.

Conditions for Dephosphorization

Dephosphorization requires oxidizing and basic slag:

$$2[P]+5(FeO)+3(CaO)=(3\ CaO.P_2O_5)$$

$$K_9=\frac{a_{(3\,CaO.P_2O_5)}}{\left(a_{(CaO)}\right)^3\left(a_{(FeO)}\right)^5\left(a_{[P]}\right)^2}$$

$$\frac{a_{(3\,CaO.P_2O_5)}}{\left(a_{[P]}\right)^2}\approx\frac{(wt\%p)}{[wt\%p]}=K_9\times\left(a_{(CaO)}\right)^3\left(a_{(FeO)}\right)^5$$

- a_{CaO} in slag should be high. This means slag should have free dissolved lime. High basicity of slag is required.
- a_{FeO} in slag should be high; slag should be oxidizing. However for efficient dephosphorization the FeO content of slag should be in between 15 to 16%.
- Low temperature favours high K_9

Conditions for Simultaneous Removal of C and P

Removal of C and P both require oxidizing conditions but P removal is possible only when a basic and limy slag is formed. Consider the following reactions occurring simultaneously

$$[C]+[O]=\{CO\}$$
$$2[P]+5[O]=(P_2O_5)$$
$$K_{CO}=\frac{pco}{[wt\%C][wt\%O]}$$
$$K_p=\frac{Np_2O_5\gamma p_2O_5}{[wt\%P]^2[wt\%O]^5}$$

It is assumed in eq. above that henrian activity is equal to (wt %). Both reactions earlier require oxygen but one reaction requires a slag which is basic in nature in addition to oxygen. Thus, if carbon and phosphorus are to be removed simulataneously, an important requirement is the availability of slag which acts as a sink for (P_2O_5). Thermodynamically slag is required in which activity coefficient of P_2O_5 is very low. The question is how low activity of P_2O_5 should be. This value can be determined by equations above Replacing [wt% O] in equation above, and after rearrangement,

$$\gamma_{P_2O_5}=\frac{K_p[wt\%P]^2}{K_{CO}{}^5[wt\%C]^5N_{P_2O_5}}$$

From equation above one can determine the value of activity coefficient of P_2O_5 which can lead to simultaneous removal of carbon and phosphorus.

Let us calculate the $\gamma_{P_2O_5}$ when molten metal contains 2% C and 0.15% P and temperature T=1773K. The mole fraction of P_2O_5 in slag is 0.1.

$$K_{co}=\frac{1305}{T}+1.979$$

We can calculate K_p and K_{co} from equations above. Substituting all the values into equation 16 we get $\gamma_{P_2O_5}$. We can also calculate $\gamma_{P_2O_5}$ at temperatures 1673K and 1873K.

The results are given below

T(K)	$\gamma_{P_2O_5}$
1673	1.6×10^{-21}
1773	1.32×10^{-22}
1873	1.42×10^{-23}

The calculations show:

- Both decarburization and dephosphorization are possible simultaneously in presence of slag in which $\gamma_{P_2O_5}$ has extremely low value.
- Low temperature requires $\gamma_{P_2O_5}$ in slag to be higher than at high temperature.

  Thus low temperature is favourable.

Deoxidization

Deoxidization is a method used in metallurgy to remove the oxygen content during steel manufacturing. In contrast, antioxidants are used for stabilization, such as in the storage of food. Deoxidation is important in the steelmaking process as oxygen is often detrimental to the quality of steel produced. Deoxidization is mainly achieved by adding a separate chemical species to neutralize the effects of oxygen or by directly removing the oxygen.

Oxidation

Oxidation is the process of an element losing electrons. For example, iron will transfer two of its electrons to oxygen, forming an oxide. This occurs all throughout as an unintended part of the steelmaking process.

Oxygen blowing is a method of steelmaking where oxygen is blown through pig iron to lower the carbon content. Oxygen forms oxides with the unwanted elements, such as carbon, silicon, phosphorus, and manganese, which appear from various states of the manufacturing process. These oxides will float to the top of the steel pool and remove themselves from the pig iron. However, some of the oxygen will also react with the iron itself.

Due to the high temperatures involved in smelting itself, oxygen in the air may dissolve into the molten iron while it is being poured. Slag, a byproduct left over after the smelting process, is used to further absorb impurities such as sulfur or oxides and protect steel from further oxidation. However, it can still be responsible for some oxidation.

Some processes, while still able to lead to oxidation, are not relevant to the oxygen content of steel during its manufacture. For example, rust is a red iron oxide that forms when the iron in steel reacts with the oxygen or water in the air. This usually only occurs once the steel has been in use for varying degrees of time. Some physical components of the steelmaking process itself, such as the electric arc furnace, may also wear down and oxidize. This issue is typically dealt with by the use of refractory metals, which are resistant to change.

If steel is not properly deoxidized, it will have reduced various properties such as tensile strength, ductility, toughness, weldability, polishability, and machinability. This is due to forming non-metallic inclusions and gas pores, bubbles of gas that get trapped during the solidification process of steel.

Types of Deoxidizers

Metallic Deoxidizers

This method of deoxidization involves adding specific metals into the steel. These metals will react with the unwanted oxygen, forming a strong oxide that, compared to pure oxygen, will reduce the steel's strength and qualities by a lesser amount.

The chemical equation for deoxidization is represented by:

$$nD + mO \rightarrow D_nO_m$$

where n and m are coefficients, D is the deoxidizing agent, and O is oxygen.

Thus, the chemical equilibrium equation involved is:

$$K_{eq} = a_{ox} / (a_D^n * a_O^m)$$

where a_{ox} is the activity, or concentration, of the oxide in the steel, a_D is the activity of the deoxidizing agent, and a_O is the activity of the oxygen.

An increase in the equilibrium constant K_{eq} will cause an increase in a_{ox}, and thus more of the oxide product.

K_{eq} can be manipulated by the steel temperature via the following equation:

$$\log K_{eq} = A_D / T - B_D$$

where A_D and B_D are parameters specific to different deoxidizers and T is the temperature in K°. Below are the values for certain deoxidizers at a temperature of 1873 K°.

Deoxidizer	**A**	**B**	**K_{eq}**
Manganese	12,440	5.33	1.318
Silicon	30,000	11.5	4.518
Aluminum	62,780	20.5	13.018

Below is a list of commonly used metallic deoxidizers:

- Ferrosilicon, ferromanganese, calcium silicide - used in steelmaking in production of carbon steels, stainless steels, and other ferrous alloys
- Manganese - used in steelmaking
- Silicon carbide, calcium carbide - used as ladle deoxidizer in steel production
- Aluminum dross - used to deoxidize slag in secondary steelmaking
- Calcium - used as a deoxidizer, desulfurizer, or decarbonizer for ferrous and nonferrous alloys
- Titanium - used as a deoxidizer for steels
- Phosphorus, copper(I) phosphide - used in production of oxygen-free copper
- Calcium hexaboride - used in production of oxygen-free copper, yields higher conductivity copper than phosphorus-deoxidized
- Yttrium - used to deoxidize vanadium and other non-ferrous metals
- Zirconium

- Magnesium
- Carbon
- Tungsten

Vacuum Deoxidation

Vacuum deoxidation is a method which involves using a vacuum to remove impurities. A portion of the carbon and oxygen in steel will react, forming carbon monoxide. CO gas will float up to the top of the liquid steel and be removed by a vacuum system.

As the chemical reaction involved in vacuum deoxidation is:

$$C + O \rightarrow CO$$

the reaction between carbon and oxygen is represented by the following chemical equilibrium equation:

$$K_{CO} = P_{CO} / (a_C * a_O)$$

where P_{CO} is the partial pressure of the carbon monoxide formed.

Decreasing the oxygen activity(a_O) will result in a higher equilibrium constant, thus more product, CO. To achieve this, subjecting the pool of steel to vacuum treatment decreases the value of P_{CO}, allowing for more CO gas to be produced.

Diffusion Deoxidation

This method relies on the idea that deoxidation of slag will lead to the deoxidation of steel.

The chemical equilibrium equation used for this process is:

$$K_{FeO} = a_{[O]} / a_{(O)}$$

where $a_{[O]}$ is the activity of the oxygen in the slag, and $a_{(O)}$ is the activity of oxygen in the steel.

Reducing the activity in the slag ($a_{[O]}$) will lower the oxygen levels in the slag. Afterwards, oxygen will diffuse from the steel into the lesser concentrated slag. This method is done by using deoxidizing agents on the slag, such as coke or silicon. As these agents do not come into direct contact with the steel, non-metallic inclusions will not form in the steel itself.

Principles of Deoxidation

Refining of hot metal to steel is done under oxidizing atmosphere. During refining oxygen dissolves in steel. Solubility of oxygen in steel is negligibly small. During solidification of molten steel, excess oxygen is rejected by the solidifying steel. This excess oxygen produces defects like blow holes and non- metallic oxide inclusion in solidified casting. Defects have considerable effect on mechanical properties of steel.

Therefore, it is necessary to remove oxygen from steel; removal of oxygen is called deoxidation.

Sources of Oxygen in Steel

- Rust on steel
- Oxygen blowing
- Steelmaking slag
- Atmospheric oxygen dissolved in steel during teeming
- Oxidizing refractories

At $1600^{\circ}C$ solubility of oxygen in liquid steel is 0.23% which deceases to 0.003% in solid steel during solidification. According to the degree of deoxidation, carbon steels may be subdivided into three groups:

1. Killed steel: Oxygen is removed completely. Solidification of such steels does not give gas porosity (blow holes).
2. Semi –killed steel: Incompletely deoxidized steels containing some amount of oxygen which froms CO during solidification.
3. Rimming steel: Partially deoxidized or non- deoxidized low carbon steels evolving sufficient CO during solidification. These steels have good surface finish.

Deoxidation of Steel

Deoxidation can be carried out either by single element such as Si, Al, Mn etc or by mixture of elements such as $Si + Mn$, $Ca - Si - Al$ etc. Deoxidation by single element is known as simple deoxidation, whereas deoxidation by a mixture of elements is known as complex deoxidation. In both simple and complex deoxidation, oxide is formed; hence it is also termed precipitation deoxidation. Deoxidation is also carried out by carbon under vacuum; which is called vacuum deoxidation. Elements are added in the form of Ferro-alloys $Fe - Si$, $Fe - Mn$ or $Fe\,Si + Fe\,Mn$ etc.

Simple deoxidation can be represented by

$$a[M] + b[O] = (M_a\,O_b)$$

If deoxidation product is pure then activity of $M_a\,O_b = 1$ and if elements are in dilute solution

$$[W_M]^a [W_O]^b = Km.$$

Where K_M is deoxidation constant and equals to $\frac{1}{K_M^1}$ where K_M^1 is equilibrium constant.

$$\log K_M = -\frac{X}{T} + Y$$

Where X and Y are constants and T is temperature. Increase in T increases K_M. Using equations above one can calculate the variation of W_O with W_M when W_M is in small quantity. For larger W_M, interaction parameters need to be considered.

$$\log f_M = \Sigma_J\, e_M^J\, W_j$$

$$\log f_o = \Sigma_J\, e_o^J\, W_j$$

Where J denotes all alloying elements. If steel contains C, and Mn.

$$\log f_o = e_o^o W_o + e_o^c W_c + e_o^{Mn} W_{Mn}$$

Where e is Interaction Parameter

All oxide products are definite compounds except compounds formed by Mn deoxidation. Here product is either solid MnO or liquid FeO. MnO.

In complex deoxidation where a mixture of Si + Mn, Ca + Si, Ca + Si + Al is used, the following advantages are reported as compared with simple one:

- The dissolved oxygen is lower.
- Due to formation of liquid deoxidation product agglomeration of the product into large size can be obtained easily and can be floated easily.

According to equation earlier

$$[W_o]^b = \frac{K_M}{[W_M]^a}$$

Equation above indicates that weight percent oxygen in steel depends on value of K_M for small concentration of deoxidizers. At 1600°C the value of Km is 2.4×10^{-5} for the reaction

$Si + O_2 = SiO_2$ and for the reaction

$2\,Al + 3O = Al_2O_3$

The value of $K_M = 3.32 \times 10^{-14}$. Similarly for the reaction $Ca + O = CaO, K_M = 9.84 \times 10^{-11}$. The value of K_M indicates the deoxidizing ability of an element. For the above reaction, calcium is the most efficient deoxidizer and Si is not so efficient as compared to calcium. Aluminum is also a strong deoxidizing element when compared with silicon.

Though calcium and aluminum are very efficient deoxidizers, but they oxidize very fast and moreover, their density is much lower than steel. Also Ca has a boiling point 1485°C which means calcium is gaseous phase at the steelmaking temperature. Suitable injection methods or addition methods are to be developed.

Kinetics of Deoxidation

Total oxygen in steel equals to dissolved oxygen + oxygen present in deoxidation products SiO_2, Al_2O_3, MnO etc). Even if the dissolved oxygen is low, deoxidation products (also called inclusions) have to be removed, otherwise steel is not clean (clean steel refers to number and size of inclusions in steel). Kinetics of inclusion is concerned with deoxidation reaction and separation of deoxidation products as well. The deoxidation process consists of the following steps:

- Dissolution and homogenization of de oxidizer. Mechanism of dissolution depends on melting point. Ferro alloys melt at around $1500^{\circ}C$. Aluminium is expected to melt faster due to its much lower point. Intensity of agitation will govern the homogenization of deoxidizer in steel melt for faster kinetics of reaction between oxygen and deoxidizer.
- Nucleation of solid product becomes easier if interface is present. Deoxidation by Al produces solid Al_2O_3 and as $Al_2O_3/steel$ such interface is useful for nucleation.
- Growth of the deoxidation product: It depends on the state of the product. A liquid product can coalesce easily as compared with the solid product. Deoxidation with single elements like Al, Si etc. produce solid deoxidation product at the steelmaking temperature. Deoxidation with ferro silicon+ ferro manganese produces liquid deoxidation product. Boron, titanium zirconium are also quite effective deoxidizers. Manganese and silicon are used in the ratio 7:1 to 4:1 in order to obtain a thin liquid slag.
- Removal of deoxidation product: Removal of deoxidation product is equally important. It is achieved by floatation and absorption into a slag. Following steps are important for removal of de oxidation products from steel:

  i) Growth of deoxidation product

  ii) Movement through molten steel to surface

  iii) Absorption of inclusion by a suitable designed synthetic slag.

Floatation of an oxide product depends among physical properties of steel, on the size of the product. The rate of rise of a spherical particle in a quiet fluid or in a fluid of laminar flow (i.e. at very low Reynold's number) can be described at steady state by Stoke's law:

$$V_t = \frac{gd^2\Delta p}{18\eta}$$

V_t = terminal velocity (m/s) of the inclusion, g is acceleration due to gravity, (m/s^2), $\Delta\rho =$ difference in density of steel and deoxidation product and η is viscosity of steel $(kg/m.s)$.

Deoxidation products are lighter than steel; hence they move up. According to equation 8 the rising velocity is proportional to square of the size of the deoxidation product. Larger sizes move faster. Moreover different sizes of deoxidation product will move with different velocities. During their movement, they may collide with one another. Stirring of melt may help floating of de oxidation products.

Degree of stirring in the melt is important. Vigorous stirring may not be of much help since deoxidation product may be circulated in the liquid.

For the removal of deoxidation product, equally important is the design of synthetic slag to absorb the deoxidation product.

Deoxidation Practice

Deoxidation can be carried out during tapping, in ladles runners and even in moulds. Bath stirring is important. During tapping, bath is stirred due to potential energy but this subsides towards the end. When deoxidation is carried out in ladle, it is called ladle deoxidation in industrial practice. Depending on the extent of deoxidation, killed, semi killed and rimming steels are produced. For carbon content less than 0.15% and enough oxygen in steel, rimming steel can be produced.

Alloy steels are fully killed to obtain maximum recovery of alloying additions.

Illustration

Let us take an example of deoxidation of steel with ferromanganese. Manganese is a weak deoxidizer. We intend to reduce the dissolved oxygen in steel from 0.045 wt.% to 0.018 wt.%. How much manganese would be required?

Consider the Reaction

$$(\mathrm{MnO}) = [\mathrm{Mn}] + [\mathrm{O}];\ \log K_{Mn} = -\frac{11070}{T} + 4.526$$

We can write $\frac{W_{Mn}}{X_{MnO}} = \frac{K_{Mn}}{W_O}$, we have assumed $h_{Mn} = W_{Mn}$ and $h_O = W_O$, where Wo is weight percent

Substituting the value of $K_{Mn} = 0.0413$ 1t $1600^\circ C$ and $W_o = 0.018$ we get

$$\frac{W_{Mn}}{X_{MnO}} = 2.29$$

For the deoxidation with manganese, the following reaction must also be considered

$$(\mathrm{MnO}) + [\mathrm{Fe}] = [\mathrm{Mn}] + (\mathrm{FeO});\ \log K_{Mn-Fe} = -\frac{9370}{T} + 4.330$$

By writing equilibrium constant and using the value of $\frac{W_{Mn}}{X_{MnO}} = 2.29$ we can find $X_{FeO} \approx 0.9$ and $W_{Mn} = 2.061$ wt. percent.

Total manganese required would be equal to manganese required to remove (0.045% – 0.018%) oxygen + Manganese required to raise the content of mn from 0.1% to 2.061%

Calculation gives 20.53Kg Mn/ton steel. If the manganese content of ferromanganese is 60%, ferromanganese would be 34 kg/ton of steel.

References

- Hudlický, Miloš (1996). Reductions in Organic Chemistry. Washington, D.C.: American Chemical Society. p. 429. ISBN 0-8412-3344-6
- Hudlický, Miloš (1990). Oxidations in Organic Chemistry. Washington, D.C.: American Chemical Society. p. 456. ISBN 0-8412-1780-7
- Phillips, John; Strozak, Victor; Wistrom, Cheryl (2000). Chemistry: Concepts and Applications. Glencoe McGraw-Hill. p. 558. ISBN 978-0-02-828210-7
- Rodgers, Glen (2012). Descriptive Inorganic, Coordination, and Solid-State Chemistry. Brooks/Cole, Cengage Learning. p. 330. ISBN 978-0-8400-6846-0
- Robertson, William (2010). More Chemistry Basics. National Science Teachers Association. p. 82. ISBN 978-1-936137-74-9

Permissions

All chapters in this book are published with permission under the Creative Commons Attribution Share Alike License or equivalent. Every chapter published in this book has been scrutinized by our experts. Their significance has been extensively debated. The topics covered herein carry significant information for a comprehensive understanding. They may even be implemented as practical applications or may be referred to as a beginning point for further studies.

We would like to thank the editorial team for lending their expertise to make the book truly unique. They have played a crucial role in the development of this book. Without their invaluable contributions this book wouldn't have been possible. They have made vital efforts to compile up to date information on the varied aspects of this subject to make this book a valuable addition to the collection of many professionals and students.

This book was conceptualized with the vision of imparting up-to-date and integrated information in this field. To ensure the same, a matchless editorial board was set up. Every individual on the board went through rigorous rounds of assessment to prove their worth. After which they invested a large part of their time researching and compiling the most relevant data for our readers.

The editorial board has been involved in producing this book since its inception. They have spent rigorous hours researching and exploring the diverse topics which have resulted in the successful publishing of this book. They have passed on their knowledge of decades through this book. To expedite this challenging task, the publisher supported the team at every step. A small team of assistant editors was also appointed to further simplify the editing procedure and attain best results for the readers.

Apart from the editorial board, the designing team has also invested a significant amount of their time in understanding the subject and creating the most relevant covers. They scrutinized every image to scout for the most suitable representation of the subject and create an appropriate cover for the book.

The publishing team has been an ardent support to the editorial, designing and production team. Their endless efforts to recruit the best for this project, has resulted in the accomplishment of this book. They are a veteran in the field of academics and their pool of knowledge is as vast as their experience in printing. Their expertise and guidance has proved useful at every step. Their uncompromising quality standards have made this book an exceptional effort. Their encouragement from time to time has been an inspiration for everyone.

The publisher and the editorial board hope that this book will prove to be a valuable piece of knowledge for students, practitioners and scholars across the globe.

Index